# Your Calf

## A Kid's Guide to Raising and Showing Beef and Dairy Calves

**Heather Smith Thomas**

Storey Publishing

*The mission of Storey Publishing*
*is to serve our customers by publishing practical information*
*that encourages personal independence in harmony with the environment.*

Edited by Lorin Driggs
Cover and text design by Carol J. Jessop, Black Trout Design
Cover photograph by A. Blake Gardner
Cover and text production by Wanda Harper Joyce and Susan Bernier
Photographs by Lorin Driggs and Heather Smith Thomas
Line drawings by Jeffrey Domm, except for pages 143–144 by Paula Savastano and
    Cathy Baker and page 22 (top), 109, 111, and 112 (top) by Elayne Sears.
Technical review by Cal Chunglo
Indexed by Indexes and Knowledge Maps

*Thanks to the U.S. Beef Breeds Council and Purina Mills for*
*permission to reprint their artwork on pages 49–53 and 132–134.*

Pictured on cover:  Amanda Bardin

Printed in the United States by Capital City Press

20  19  18  17  16  15  14  13  12  11  10  9

**Library of Congress Cataloging-in-Publication Data**

Thomas, Heather Smith, 1944-
      Your calf : a kid's guide to raising and showing beef and dairy calves / Heather Smith Thomas.
         p.    cm.
      Includes bibliographical references (p.    ) and index. Summary: Offers information and advice on selecting, feeding, housing, caring for, breeding, and showing both beef and dairy cattle.
      ISBN 0-88266-947-8 (pbk. : alk. paper)
      1. Calves—Handbooks, manuals, etc.—Juvenile literature [1. Cattle—Handbooks, manuals, etc.]
I. Title
SF205.T48    1996
636.2'07—dc20
                                                                        96-32097
                                                                            CIP
                                                                            AC

# contents

# Dedication

This book is dedicated to "Norman." She was a tiny heifer calf who spent the first four weeks of her life in our kitchen in a big cardboard box because the subzero temperatures outside were much too cold for her frail little body. She stayed in the house until she was big enough to go outside. We lifted her out of her box every day for exercise, and she loved to gallop around the kitchen, sometimes slithering and sliding on the slippery linoleum floor like little Bambi on the ice pond.

Norman grew up to be a cow. She is still a special pet, and seems to enjoy her privileged status in our cow herd.

# A Few Words About This Book

*Y*our Calf will be a handy reference for any youngster who wants to raise a calf or learn more about cattle — from the young beginner with a bucket calf, to the high school student with a growing herd of cows as a "college fund" investment. This book is written in simple terms so even the beginner or young stockman can understand it. At the same time, it covers a wealth of information that will be useful to any young person with an advanced beef or dairy project. It will also be a help to parents of kids with 4-H or FFA (Future Farmers of America) projects, and can be a long-lasting "cattle manual" that the child can continually refer to as he progresses farther with his projects or raises cows of his own.

For the child desiring to raise a calf, I would recommend joining a 4-H club or FFA chapter at school. If there are none where you live, you'll want to get advice from time to time from a veterinarian, cattle breeder or dairyman, or your county Extension Service agent. Don't be afraid to contact an experienced person to ask questions or to request help.

# Introduction

Most kids love animals. The responsibility of caring for an animal helps build confidence in your own abilities, and gives a feeling of satisfaction and self-worth. A cat or dog is nice, and so are rabbits and chickens. But maybe you want to tackle something more.

## Why Raise a Calf?

Raising a calf is challenging and fun, and sometimes helps chart a path for the future. You might like to be a farmer or rancher someday, or you might simply be interested in cattle. Maybe you imagine how exciting it will be taking care of and working with an animal that will become much bigger than you are.

Raising a calf is not only fun, but can also earn you money. You can sell your calf for beef, or you can raise a heifer to be a cow, and have calves of her own. Many kids have earned money for college by raising calves.

## What Kind of Calf Should I Raise?

You have several choices when it comes to deciding what kind of calf to raise. Your choices will depend on how much space you have and what your goals are.

**Heifer.** *A young female cow that has not yet had a calf.*

## Space

Do you want to raise a steer to sell for beef, or a heifer that will grow up to be a cow? The decision may depend in part on how much space you have. A calf can be raised in a small area, even in your backyard, if it's not in town where city zoning ordinances forbid livestock. But if you are going to have a big cow that someday will have a calf of her own, she'll need more room. You may want to start with a calf to raise for beef, and later try a heifer project if you can find a larger place to keep her when she grows up.

## Goals

The kind of calf you get also depends on your goals.

If you are raising a beef calf to earn money, you should probably raise a steer. Steers bring more money per pound when sold, and also weigh more than heifers of the same age. But if you want to keep your calf, you should choose a heifer. Or you may want to raise a dairy heifer to sell. Dairy heifers are worth more money than beef cattle when they grow up.

If you want to raise a heifer, do you want a beef cow or a dairy cow? Do you want a cow that will have calves that grow up to be sold for beef, or a cow you can milk or sell as a dairy cow?

## Steer or Heifer?

A steer is a male calf that has been castrated — his testes or testicles have been removed. A steer you raise will end up as beef. You will sell him at the end of your project, or whenever he grows up. A heifer is a female calf. A heifer can be raised for beef, or she can be kept to become a cow.

If you have no experience with cattle, you may need to learn how to tell the difference between a steer and a heifer. It's sometimes difficult to tell, unless you know what to look for.

## Recognizing Males

Bulls and steers are males. When a male calf is born, he is a bull. His male reproductive organs are his testicles and his penis. His testicles are inside his scrotum — the saclike pouch that hangs between his hind legs. His penis is inside his sheath, located on the underside of his belly. He urinates from his penis.

Most bulls end up as steers. Only the best males are kept as bulls for breeding with cows. A commercial herd may include no bull calves. The rancher may buy all his bulls from a purebred breeder. Beef calves are sold as steers.

## Becoming a Steer

A bull becomes a steer when his testicles are removed in a process called "castration." The steer may still have a scrotum or his scrotum may be entirely gone, depending on the method used to castrate him. (See pages 122–123.) If he has a scrotum it is much smaller than a bull's scrotum, because the testicles are no longer in it.

## Recognizing Females

A heifer is a young female cow. Most of the heifer's reproductive organs are inside her body, so you can't see them. She has a tiny udder between her hind legs with little teats on it. A bull or steer calf also has small teats, just as a boy has nipples, but they don't grow large.

The heifer urinates from an opening by her vulva, which is located under her tail, below the rectal

**Bull.** *A male bovine that has not been castrated.*

Bull

Steer

Heifer

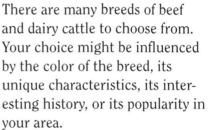

Cow

*Breed. A group of animals with the same ancestry and physical characteristics.*

opening. The vulva is the opening to the reproductive organs located inside her body.

## Becoming a Cow

The female is called a heifer until she is older than two years and has had a calf. Then she becomes a cow.

## What Breed Should I Choose?

There are many breeds of beef and dairy cattle to choose from. Your choice might be influenced by the color of the breed, its unique characteristics, its interesting history, or its popularity in your area.

Once you decide on the breed you want, have a knowledgeable person help you pick out a calf. There are some outstanding cattle in every breed. Be sure you get a good calf that will grow fast and do well for you at the fair or become a good cow.

## Will I Need Help?

Raising a calf involves work. You will have to feed and care for the calf every day on a regular schedule. You cannot skip chores on days you might want to do something else.

Some aspects of care may require assistance from your parents, an older brother or sister, or a friend or acquaintance who is familiar with cattle. Most things, however, you will be able to do yourself, especially if you are starting with a baby calf who is not already bigger and stronger than you are.

## Using This Book

This book explains what is involved in keeping and raising a calf.

Part One contains important information that will help you regardless of the kind of calf you decide to raise. Part Two is about raising a beef calf. There are also chapters on keeping a beef heifer to become a cow. Part Three is about raising a dairy heifer.

A few things are the same with all calves, but there are differences in raising a beef calf (which you'll probably buy as a big weaned calf) and a dairy heifer (which you'll probably buy as a new baby calf).

## Good Luck, and Have Fun!

Whether you raise one calf or many, a beef calf or dairy calf, you'll find the experience rewarding and fascinating. The best part is getting to know your animals as individuals. They all have unique personalities.

## Cattle and Humans

Humans have been working with cattle since prehistoric times. First we hunted them. Wild cattle were the main meat in the diet of Stone Age people.

Some of our prehistoric ancestors captured cattle to tame about 10,000 years ago. They wanted a handy supply of meat for food and hides for leather clothing. After taming the cattle, humans discovered they could hitch these animals to a cart or a plow. Oxen were used for transportation before horses were. Cattle were domesticated before horses, probably because they were easier to catch.

# PART ONE

# Basic Cattle Care

# Finances

**R**aising a calf is a lot of fun and also a challenge. At the beginning, you will have to figure out how you are going to pay for your calf and for its feed.

## Buying and Selling

A weaned beef calf and its feed for a year will cost a lot. A newborn dairy calf won't cost as much as a weaned beef calf. A registered purebred heifer may cost more than an unregistered calf. A beef steer costs more than a heifer. A dairy heifer costs more than a dairy steer. The price will vary, since the cattle market goes up and down. When prices are low, you might be able to get a beef calf for 50¢ a pound. When prices go up, you might have to pay 90¢ or more. For a 500-pound steer, that's a price range of $250 to $450 or more.

## Borrowing the Money

*Interest. Money paid for the use of borrowed money.*

If you don't have savings for buying a calf, you might borrow the money from your parents or another family member. If you are in 4-H, your county 4-H leaders may have a program to help finance beef or dairy projects. In these situations, often the interest may not be due until your project is complete and your calf is sold. Sometimes a local bank will finance a beginning 4-H project, or an interested individual in your community may lend money for 4-H or FFA projects.

If you borrow from parents or a family member, it's a good idea (and good experience) to pay interest on the money, just as you would if borrowing from a bank.

If you borrow from a bank, your parents will go with you to set up the loan. They may need to sign the note also. List your proposed expenses and your expected income from the sale of your calf or from your summer job. Make a schedule for payments on the loan. Your banker will need to know: (1) How much money you need; (2) How long you will need it; and (3) How you will repay the loan if your calf dies.

## Buying on Credit

The farmer or rancher who sells you the calf may sell it on credit. You can pay him back with interest when you finish the project. You may be able to work for him on weekends or during the summer to pay for the calf.

You also need to figure out how to pay for feed. Some feed stores will extend credit, letting you pay for feed and supplies after you sell the animal.

## Estimating Costs for a Beef Project

Look at the sample budget on the next page. The initial cost of your steer might be less if calf prices are in a down cycle, or more if prices are up. The price of barley could also be more or less. Hay costs fluctuate, as well. If you raise a calf on pasture, you won't have as much expense for hay and grain. Your sample budget is just an estimate until you find out what the actual costs will be.

Additional costs might include materials for building a pen, or equipment such as feed and water tubs. You may also have transportation fees, entry fees for the fair, and veterinary expenses and supplies. If the steer in this example weighed 1,250 pounds by an August or September show and yearling steers were worth 70¢ a pound or more, you would make a profit unless your expenses were high.

## Insurance

When starting your calf project, think about insurance. You pay a small fee for the insurance. If your calf dies, the insurance company pays you the value of the calf, and you can use that money to repay the bank loan. Ask your county agent about insurance companies that do this.

## Sample Budget

Here is a sample budget — an estimate of anticipated costs for a beef project.

*Cost of steer:*
500 pounds at 80¢ per pound = $400

*Rolled barley:*
5,000 pounds (2.5 tons) at $52 per ton = $130

*Hay (grass/alfalfa mix):*
1.5 tons at $70 per ton = $105

*Total feed cost:*
$235

*Cost of steer plus feed:*
$635

## Keep Records

Financial records are an important part of your project. Good records show where you started and document how you are doing. They also give you experience in the financial aspect of raising a market animal or starting a breeding herd. If you are in 4-H, your completed record book gives a history of your financial progress.

## Costs of Raising a Calf on Pasture

Most market calves are not raised on grain and sold as show calves. They are raised by farmers and ranchers on pasture grass in the spring and summer and maybe cornstalks (after the corn is harvested) or grain stubble (after the wheat, oats, or barley is harvested) in the fall and winter. They graze on this "crop aftermath" for awhile to grow big inexpensively, and then go into feedlots for finishing on grain to make them fatter before butchering. Calves raised on pasture don't get big as fast as grain-fed show calves, but they cost less to feed. So they make just as much or more money.

You will still have to buy hay for the winter, and keep your calf in a pen and feed him hay until the pasture grows tall enough for him to eat in the spring. But if you have pasture, you can raise a calf quite cheaply (pasture rent is much cheaper than grain and hay) and get him big enough to sell or butcher by the next fall.

# Housing, Facilities, and Equipment

# CHAPTER 2

**Y**our calf will need shelter from the sun, wind, and rain. (A newborn dairy calf has special needs for shelter, which are described in chapter 14.) You'll also need equipment for feed and water, for cleaning up after your calf, and for leading and tying your calf.

## The Pen

You need a proper place to keep your calf. Even if you are going to raise him on pasture, you'll need a pen or corral for when you first bring him home, and also for catching and working with him.

The pen needs to be dry with good drainage. If necessary, sand can be put in the bedding area or a shady spot where the calf sleeps, to make sure it stays dry.

## Pen Fencing

Your pen or corral should be built of boards or poles, or very strong woven-wire netting. Regular barbed wire or smooth wire won't work because a calf can get through it if he tries hard enough or crashes into it.

You don't want an electric fence in your small catch pen. There are times you need to corner the calf in the pen, and you don't want him to be shocked. The pen should be a safe, secure place for your calf, where he won't have to worry about being shocked. And sometime you might get pressed against the fence or need to climb over it, and you don't want to get shocked, either.

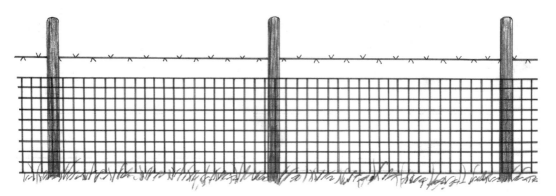

Wire fence with braces and metal stays

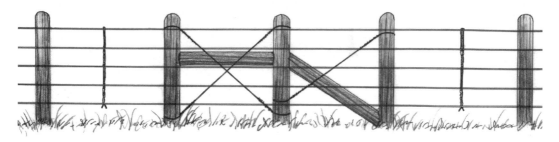

Net wire fence with one barbed wire on top

Corral fence with posts and poles

## Building a Pen or Corral

If you don't have a pen, perhaps your family can help you build one. You'll need sturdy wood posts. The posts should be long enough to set deeply into the ground and still make a fence at least 5 feet high. You'll need poles, boards, or wood or metal panels to secure to the posts. Strong woven-wire netting will also work. The posts should be close together. A good pen will cost money to build, but it will last a long time.

## How Big Should the Pen Be?

If the calf will be spending all of his time in the pen, it should be large enough to give him room for exercise. To be adequate, the pen should have at least 900 to 1,000 square feet. This can be provided by a variety of dimensions such as 10-feet-by-100-feet, 20-feet-by-50-feet, or 30-feet-by-30-feet — whatever works to fit the space you have. If you are raising more than one calf,

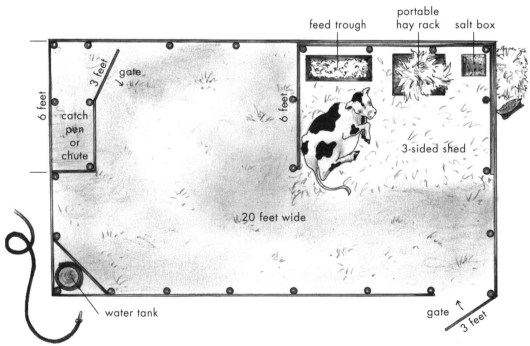

Typical layout for a calf pen and shed

you should add at least 200 to 300 more square feet of space to the total area for every additional calf. There should be shade in the pen, from a building or a tree. The calf needs about 100 square feet (an area 10-feet-by-10-feet) of summer shade.

### Digging Postholes and Setting Posts

A pen should be built on solid ground. Posts set in a boggy, wet area will become loose and wobbly. The postholes can be dug with just a shovel if the ground is mainly dirt with few rocks. But if it is rocky, you will need an iron bar to jar the rocks loose as you dig.

Make sure you set the posts in a straight line. A crooked fence will not be as strong. Set your corner posts and sight between them to line up your postholes and your posts, or stretch a long string between them to give you an exact line. The posts should be set at least 2½ feet in the ground. They should be treated on the bottom end with wood preservative so they won't rot. The preservative should cover all of the portion set in the ground, and extend a few inches above ground level.

When filling in the dirt and rocks around a post, put in a little at a time, tamping it firmly with your iron bar or a tamping stick before putting in any more dirt. The secret to having tight, solid posts is to set them deep enough and to tamp the dirt very firmly around them.

### Catch Pen or Chute

If you buy a calf that is not yet accustomed to people, you need a small catch pen in one corner of your pen. You also need a place where you can restrain the calf for giving shots and medications. A small enclosed shed or feeding area can be used for cornering and catching the calf. You can also make a head catcher or stanchion in his feeding area, where you can lock his head in place when he sticks it through to eat.

You can make a little chute at one corner of the pen. Then you can herd him along the fence and swing the gate shut behind him. A swinging panel will also work if you haven't had time or materials to build a chute before you need to capture him.

## Pasture

Maybe your family has a few acres of pasture. Or maybe you know someone who would lease you some pasture. Pasture can often be rented for a small fee per month.

Always check the pasture fences carefully to make sure they are in good shape. A calf that has never lived alone may be very frantic when you first bring him home. He'll want to go back to the herd he lived with, and he may try to get out.

## Wire Fence

A good wire fence will hold cattle that are not being crowded or trying hard to get out. But the wire must be tight, not saggy, so your calf won't get into the habit of reaching through it. If he can reach through to eat grass, he may eventually crawl through.

Net wire is the best, since a calf cannot crawl through it. Carefully inspect the fence before you put a calf in the pasture. Replace any missing staples on wood posts or clips on metal posts so that the wire is attached properly.

Tighten any loose or sagging wires. Make sure there are no "holes" where a calf might be able to get out. A dry ditch can make a handy escape route; a calf can walk right under the fence. Put a pole across the ditch, under the fence, and attach it securely so the calf can't just push it away.

## Electric Fence

An electric fence will keep your calf from trying to crawl through a wire fence or rubbing on it. After he has been shocked a few times, he won't touch the fence.

You can use portable or temporary electric fences to divide a large pasture into several smaller ones for pasture rotation. Then you can let the calf graze in one part while the other parts grow taller, moving him to a new section after he eats one down.

For an electric fence you will need a battery-powered or electric fence charger and insulators to attach the wire to the fence posts. The electric wire cannot touch anything metal or it will short out and won't work. It shouldn't touch wood posts or poles, either, or it will short out whenever the wood is wet. Keep all weeds and brush clipped around the electric fence to avoid starting a fire.

## Shelter

Your calf's pen should have shelter so he can get out of the wind, rain, and hot sun. A simple shed can be made with tall posts that support a roof. A free-standing shed with walls on three sides will also give the calf protection from bad weather. Two sheets of plywood placed on each side of a fence corner make a very nice windbreak. You can add another sheet of plywood over the

A simple shed with two or three walls can provide shelter.

top to make a roof. The roof should slope so that rain or melting snow will run off. Make sure it slopes away from the main pen so that the water doesn't create mud in your pen, or flow into the shed.

The shed should be on a high, dry spot, with good drainage. Before you build a shed, figure out which way the wind usually blows. Situate the shed walls to give the greatest protection from wind.

A bale of straw, bark mulch, or wood chips scattered into a bed in this protected corner will give the calf a dry place to sleep. Make sure his bedding area is in a high, dry spot.

## Arrangements for Water and Feed

You'll need to make sure your calf has feed and water every day.

### Water Supply

Your calf will need a source of fresh water, even if that means a tub or tank that you fill twice a day with a garden hose. If you put a water tub in his shed, it should be up off the ground, but no higher than 20 inches so he can drink easily. A tub on the ground gets dirty since the calf may step in it or poop in it. You can make a stand or frame to hold the tub. A handy way to do this is to nail a board across the corner of the stall, leaving room for a tub or bucket to fit snugly. You can pull the tub or bucket up out of its "holder" to rinse and clean it, but the calf can't tip it over.

The calf will drink more in hot weather than in cold weather. Make sure he always has enough water, and that it doesn't freeze in winter. You may have to break ice every morning and evening. A rubber tub is handy in winter since it can be tipped over and pounded on to get the ice out without developing a leak.

## Water Trough Idea

A water trough can be made from anything that will hold water and can be cleaned easily, such as an old bathtub or washtub.

## Prevent Frozen Hoses

If you water your calf with a hose, be sure to drain it thoroughly after each use in cold weather so it doesn't freeze.

Out in the pen or pasture, an old tire can serve as a bucket holder, to keep it from spilling.

## Feed Trough

An inexpensive feed trough can be made using 2-inch lumber cut into lengths. If several calves will be using it, allow 3 square feet per calf. Make the sides of the trough at least 6 inches deep. If up off the ground, a good height is about 18 to 20 inches.

Situate the water away from the feed rack or feeding area, so your calf won't drag feed into the water. You also want it far from where he beds. If he has to walk a ways to the water, he won't be as likely to stand close to it and poop in it by accident. Keep the water fresh and clean, even if you have to dump and rinse his tub every day. If it's not clean, he won't drink it. Use a tub or tank that is easy to dump and rinse.

## Feed Rack or Feed Box

If you feed your calf hay on the ground, he will waste some. He won't eat hay that has been stepped on, is muddy, or has manure on it. If you make him a feed rack or feed manger, he won't waste so much hay. You will need to clean out the rack now and then. If he

You can build an inexpensive feed trough
from 2-inch lumber.

doesn't eat all the hay in the bottom or corners, it may mold if it gets wet. This is one reason to have the feed area inside the shed; the feed won't get wet.

If you feed your calf outdoors, you can make a little roof over his feed manger, hay rack, or grain box or tub, so moisture from rain or snow won't ruin the feed. Moldy feed is not good for your calf and might make him sick. Moldy hay is also dustier; mold spores can get into the air when he eats and may make the calf cough.

## Grain Box

If you are fattening your calf as a 4-H project or show steer, you will be feeding grain every day. You'll need a good trough or grain box, or a rubber tub on a feed stand where he can't pull it off or step in it. Whatever you put the feed in should be up off the ground so the calf won't put his feet in it. He won't eat dirty grain. A simple rubber tub works fine if you have only one calf. It's very easy to clean, and can be washed out. Put a roof over the tub or feed box, since a calf won't eat wet grain.

Clean out any leftover kernels before adding new grain. The grain should always be fresh. If birds have pooped in the tub or trough, or there's any buildup of old, fermented, or moldy grain in the corners, the calf may refuse to eat the next batch you put in.

## Manure Disposal

Your calf will create a lot of manure. Out in a large pasture this is no problem. It serves as fertilizer. But in a pen or shed, it will be concentrated in a smaller area. After awhile it will build up and need to be cleaned out. A corral may be easiest to clean with a tractor and blade or loader, but you can keep his shed or bedding area clean with a wheelbarrow and a manure fork. (A manure fork is like a pitchfork, but with more tines. The manure and straw can't fall through it so easily.)

A salt box or grain box can be made of four boards (1"x8") and a bottom (1"-thick board).

If your calf spends much time in his shed, clean out the manure and soiled bedding regularly. Then it won't build up and become a difficult job.

Manure makes excellent fertilizer. It contains just the right ingredients that plants need for growing strong and healthy. You can spread the manure over your pasture, or use it for your family's garden. You can make a compost pile with the manure and the old bedding that you clean out of your shed and corral. People in your neighborhood who have gardens might buy manure from you. This is often a good way to get rid of your compost pile and make a little extra money as well.

## Storage Space

Your calf-keeping facilities should include storage space in your barn, shed, or garage. You'll need a place to store feed and feed containers and you'll need shelves and wall hangers for tack and equipment.

## Halters and Ropes

You'll use halters and ropes to lead and tie your calf when you are working with him.

### Halters

A good halter will be one of your most important items of equipment. You can buy a calf halter, but you can also make a very good one yourself. Your 4-H leader can show you how to make an adjustable halter and lead from a 12-foot length of rope. The halter can be made larger or smaller very quickly and easily and will be useful as your calf grows.

When putting the halter on a calf, always place it so that the adjustable part is on his left side. This is the side you lead the calf from. When you pull the halter tight, pressure should be on the rope under the calf's chin, and not behind his ears.

## Tying Knots

Before you start halter training your calf, learn how to tie a good knot. You don't want him to get away when you tie him up. You need to know how to tie knots that will stay tied, and can be untied easily even if the calf has pulled hard on the rope. For your own safety and the safety of your calf, it helps to know how to tie the right kind of knot for each situation.

The end of the rope that you will be tying is called the "working end." The rest of the rope is called the "standing part."

*Overhand knot.* The easiest knot to tie is the simple overhand knot — the one you make first when tying your shoes. This knot is often the first step in forming more complex knots.

*Square knot.* This is a more useful version of the overhand knot; it is just two overhand knots, one tied on top of the other. If tied correctly, this is a perfect knot for joining two pieces of rope, such as tying a broken rope back together, or for tying a rope around a gate and gatepost to keep the gate closed. A properly tied square knot will not slip.

*Quick release knot.* This knot (also called a reefer's knot, a bowknot, or a manger tie) is useful for tying your calf to a fence post or to his stall at the fair. Like the square knot, it is a good nonslip knot. The quick release knot has the advantage of being more easily untied when it is pulled really tight, as when your calf has pulled back on his rope.

*Bowline knot.* This is probably the most useful non-slip knot when working with livestock. You can tie a rope around an animal's neck or body without danger of it tightening up when the rope is pulled. And it is relatively easy to untie.

An easy way to remember how to tie a bowline is to think of the following story. The first loop is a "rabbit hole," the standing part of the rope is a "tree," and the working end of the rope is a "rabbit." The rabbit comes out of the hole, runs around the tree, and goes back down its hole.

Overhand knot

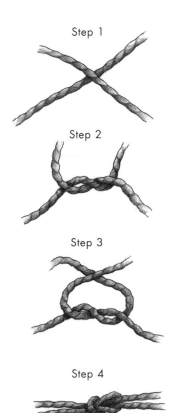

Square knot

Step 1

Step 2

Step 3

Step 4

## Double half hitch knot

**Step 1**

**Step 2**

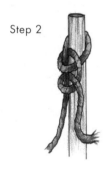

## Quick release knot

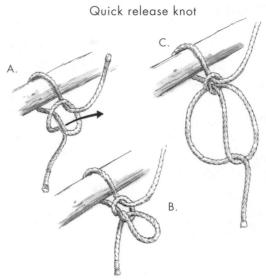

A. The end of the halter rope goes over and then around the pole. The rope goes under itself. The end of the rope is doubled into a loop.
B. The loop is put through the first loop...
C. ...and pulled through and tightened.

## Bowline knot

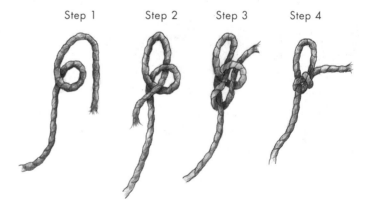

Step 1      Step 2      Step 3      Step 4

*Double half hitch.* This knot is quick and easy to tie, acts like a slipknot, and is a handy way to secure the rope around an animal's leg when tying a leg back, or to secure the end of the rope when no other knot seems appropriate.

# Nutrition

# CHAPTER 3

To feed your calf properly, you need to understand how cattle digest their food. Cattle are ruminants. This means they have four stomach compartments and chew their cud — burping up their feed and rechewing it more thoroughly. The four compartments of a ruminant's stomach are: the rumen (paunch), the reticulum (honeycomb compartment), the omasum (also called "manyplies" because of its numerous plies or folds), and the abomasum or "true stomach," which is similar to our own stomachs.

When a ruminant eats, it chews its food only enough to moisten it for swallowing. After being swallowed, the food goes into the rumen, to be softened by digestive juices. After the animal has eaten its fill, it finds a quiet place to chew its cud. It burps up a mass of food, along with some liquid, swallows the liquid part, and then chews the mass thoroughly before swallowing it and burping up some more. The re-chewed food goes into the omasum, where the liquid is squeezed out, then goes on into the abomasum.

Ruminants developed this handy way of eating so they could cram in a lot of feed at once while grazing out in the meadows, and then retreat to a safe, secluded place to chew it more fully.

## What Do Cattle Eat?

Cattle do well on a wide variety of feeds. To some extent, what you feed your calf depends on whether it

1. Rumen — storage area for feed; bacteria here help break down roughage.
2. Reticulum (honeycomb). Honeycomb-like walls catch foreign material that could injure the digestive system.
3. Omasum. Liquid is removed from the food.
4. Abomasum (true stomach). Digestive juices are secreted to finish breaking down the feed.

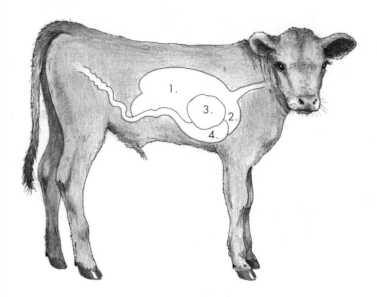

Cattle and other ruminants have four-part stomachs.

is a beef calf or a dairy calf, a steer or a heifer. Specific feeding guidelines are described in parts two and three of this book. However, the basic elements of good nutrition are the same for all cattle.

Make sure your feeds are a balanced ration — containing the basic ingredients for good nutrition: proteins, carbohydrates, fats, vitamins, and minerals.

## Proteins

Protein is necessary for growing. Protein can be supplied by high-quality legume hay such as alfalfa or clover, pasture grasses, or high-quality grass hay. Protein supplements include cottonseed meal, soybean meal, and linseed meal. A calf on good hay or pasture won't need supplements. High-quality alfalfa is a good source of protein, but care must be taken in feeding it, to avoid bloat. (See page 26.)

## Carbohydrates and Fats

Carbohydrates and fats provide energy and are used for body maintenance and weight gain. Some of the feeds that contain a lot of carbohydrates and fats are barley, wheat, corn, milo (grain sorghum), oats, and grain by-products such as millrun and molasses.

## Vitamins

Vitamins are necessary for health and growth. Green pasture, alfalfa hay, and good grass hay contain carotene, which the animal's body converts into vitamin A. Overly mature, dry hay may be deficient in vitamin A. The other vitamins needed by your calf will be in the feed or created in the calf's gut, except for vitamin D, which is formed by sunshine. Your calf will get enough vitamin D unless he spends all of his time indoors.

## Minerals

Minerals occur naturally in roughages and grain. Most cattle don't need mineral supplements beyond those found in ordinary feeds, unless your area's soils are deficient in iodine or selenium.

*Calcium and phosphorus.* Calcium and phosphorus are necessary for growth and strong bones. Calcium is well supplied by alfalfa or clover. Grains are a good source of phosphorus.

*Salt and trace minerals.* Salt is very important for your calf. All cattle need salt; it is necessary for proper body functions and also helps stimulate their appetites. It is the only mineral not found in grass or hay. Always provide salt, either in a block or as loose salt in a salt box.

Trace mineral salt can be used if your feeds are deficient in certain minerals. Trace minerals include copper, iron, iodine, cobalt, manganese, selenium, and zinc. Your vet, county agent, or 4-H leader can help you figure out what kind of salt to use and whether or not it should include trace minerals.

## Roughages and Grains

Roughages are the most natural feeds for cattle. Cattle do very well on them, but don't grow quite as fast or get as fat so quickly as when fed grain. If you are

> ### Be Careful with Supplements
>
> Check with your vet or county Extension agent before you add any supplements to your calf's diet. Some are harmful if overfed.

**Roughages.** *Feeds, such as hay and pasture, that are high in fiber and low in energy.*

raising a beef animal and not trying to push him for fast gain to get him ready for a certain show, or if you are raising a weaned heifer to keep as a cow, you will feed mostly roughages, or forage, and little or no grain. A grain-fed heifer may become too fat. But grain enables the beef calf to reach market weight or show condition faster. So his diet is generally a combination of grains and roughages.

## Roughages

Your calf will need about 3 pounds of hay daily for each 100 pounds of body weight. For example, a 500-pound calf needs about 15 pounds of good hay. After you get him started on grain, part of the hay ration can be replaced with grain.

First-cutting alfalfa often has a little grass in it and can be an ideal hay for your calf. Second – or third – cutting is generally richer and more apt to bloat the calf.

Alfalfa hay gets moldy more readily than grass hay if it gets wet or was baled too green, before it dried out. When you buy alfalfa hay, check it closely to make sure it doesn't contain mold.

## Hay Quality

Make sure any hay you buy is not moldy or dusty, since this may give your calf digestive or respiratory problems.

Hay should not be stemmy or coarse. The protein and nutrition are mainly in the leaves, so stemmy hay is not very nutritious. It's also hard for your calf to chew. When buying alfalfa make sure it is green and bright, with lots of leaves and very fine stems, not coarse, or brown and dry. Alfalfa cut early, before it blooms, is finer and more nutritious. After it blooms, it has less protein, and coarse fiber with larger stems.

### Don't Overfeed Alfalfa

Alfalfa hay is richer in vitamin A, protein, and calcium than grass hay. But it can also cause digestive problems and bloat. If the rumen gets too full of gas, it gets big and tight like a balloon, and puts pressure on the calf's lungs. He may suffocate and die. So be careful to not overfeed alfalfa. A mix of grass hay and alfalfa hay is usually safer.

## Concentrates

Grains such as corn, milo, oats, barley, and wheat are called "concentrates." In the Pacific Northwest, barley is plentiful and can be used instead of milo or corn. Wheat is usually too high priced to feed to cattle. Corn is high in energy and is commonly used for calf feed when it is available. Oats also make good feed, as does dried beet pulp with molasses.

## Pasture Management

Your calf's pasture should contain several types of grasses that will be nutritious for him. You can't expect your calf to do well in a weed patch. If you don't know anything about pasture plants, have your 4-H leader or county Extension agent look at your pasture.

### Caring for Pasture

Some pasture plants become coarse as they mature, and your calf will not eat them. Weedy areas may also be a problem. You can make the pasture better if you mow or clip these weeds so they won't go to seed and spread. If there are bare spots in the pasture, you can seed them with a "pasture mix," scattering the seeds by hand when the ground is wet.

If you live in a rainy area, the pasture will grow fine without much help. But if you live in a dry climate, the pasture won't be much good in late summer unless you keep it watered, with either a ditch or sprinklers.

Lush green grass has as much protein and vitamin A as good alfalfa hay. For a growing calf or heifer, good pasture is hard to beat. But keep a close watch on your grass. If grass gets short or dry, the calf won't grow well. Dry pasture is poor feed; it loses its protein and vitamin A. Under these conditions you should feed some good hay to help supplement the pasture.

## Selecting Hay

If you don't have any experience with hay, ask your 4-H leader, county Extension agent, or another knowledgeable person to help you select the hay for your calf.

## Let the Pasture Grow

Pen up the calf in the early spring and feed hay for awhile, to let the pasture grow. Otherwise your calf will eat the new green grass as soon as it starts and prevent it from growing tall.

# Keeping Your Calf Healthy

**J**ust like humans, cattle sometimes get sick. This chapter describes ways to prevent your calf from getting sick. It also explains how to recognize illnesses and what to do when your calf does get sick.

## Preventing Illness

Cattle need vaccinations against diseases such as blackleg, malignant edema, and brucellosis.

Talk to your vet about a vaccination program to protect your calf against the most common diseases in your area. A small herd or a single calf may not be exposed to some diseases that are more likely to affect cattle in larger groups.

There are illnesses for which there are no vaccines. Most diseases, however, can be prevented with good care and preventing exposure to contagious diseases.

## Signs of Sickness

If you get to know your calf well, and pay close attention to it every day, you will learn how to tell whether it is feeling fine or getting sick.

**Contagious.** *Readily passed from one animal to another.*

## Behavior and Appearance

The healthy calf is bright and alert, and has a good appetite. It comes eagerly to its feed, or is grazing during the times of day it usually grazes. The sick calf may spend a lot of time lying down. It may seem dull, with ears drooping instead of up and alert. It may stop chewing its cud because of pain, fever, or a digestive problem.

A calf that feels good usually stretches when it first gets up, and has an interest in its surroundings. It responds with curiosity to sounds and movement. The calf also spends some time licking itself. When it walks, it walks freely and easily.

By contrast, the sick calf may be less interested in things around it. When it does get up, it may be slowly or with an effort. All its movements are slow. It doesn't have that sparkle of vitality and health shown by a normal animal. The more serious the illness, the more indifferent the calf will be to its surroundings, and the more reluctant to move.

If the calf is overly alert or anxious, and continually looking around, this can be a sign it is in constant pain or discomfort. Some kinds of pain may make it restless — wandering about, lying down and getting up repeatedly, kicking at its belly or switching its tail, or looking around at its belly.

## Respiration Rate

Respiration rate can give a clue as to sickness or health. A sick calf with a fever will breathe fast. But remember that exercise or hot weather will also make it breathe faster. Cattle don't have many sweat glands and cool themselves by panting.

When standing quietly or lying down, your calf's respiration should be about 20 breaths per minute. An easy way to count respiration rate is by watching the calf sides move in and out. Each in-out movement counts as one respiration. Count for 15 seconds (using a second hand on a watch) and then multiply by four to get the breaths per minute.

## Types of Diseases

Disease is any change from normal. Diseases can be infectious or noninfectious. Noninfectious diseases include bloat, poisoning, founder, or injury. Infectious diseases are caused by microorganisms ("germs") or parasites. The most common microorganisms are bacteria, viruses, and fungi. Infectious diseases may be contagious (such as pinkeye) or noncontagious (such as infection of a wound or abscess).

## A Warning Sign

If you have several calves, any calf that is off by itself should be checked on. A sick animal often leaves the herd.

## Temperature

Normal temperature for a calf is 101.5°F. Anything over 102.5°F is a fever. Learn how to take your calf's temperature using a rectal animal thermometer. You can buy one from your vet. Tie a string to the ring end so you'll never lose it in the calf's rectum.

Begin with your calf tied up or in a chute. Shake the thermometer down below 98°F and moisten it with petroleum jelly so it will slide in easily. Gently lift your calf's tail and insert the thermometer into the rectum. Keep track of it. The calf may poop it out in a bunch of manure, so hold onto the string just in case. Leave the thermometer in for two minutes to get an accurate reading. When you take it out, wipe it with a paper towel so you can read it.

## Other Signs of Illness

Another clue about the health of your calf is whether its eating habits are normal. Does it chew and swallow properly, or is swallowing painful? Is it drooling, dribbling food from its mouth, or having trouble belching and chewing its cud?

Are your calf's bowel movements and urination normal? With some digestive problems the calf becomes constipated. Manure may become firm and dry, or even absent if there is a gut blockage. Or the sick calf may have diarrhea. Manure should be moderately firm (not runny and watery), and brown or green.

Urination may become difficult if there is blockage of the urinary tract, such as when a steer develops a bladder stone. The steer may dribble only small amounts of urine, or remain in the urinating position for a long time, or kick at his belly in pain, or stand very stretched. If he shows any of these signs, call your vet.

Pay attention to abnormal posture. Resting a leg may mean lameness. Arching the back with all four legs bunched together is usually a sign of pain. A bloated calf may stand with its front legs uphill to make it easier to belch gas. A sick calf may lie with its head tucked around toward its flank.

# Common Health Problems

Good care can help prevent many health problems. If your calf does get sick, recognizing the symptoms early and knowing what to do can make all the difference.

The two worst problems in calves are scours and pneumonia.

## Scours

Scours is the most common disease of young calves, and causes the most deaths. It won't be a problem in a big weaned beef calf, but could be life-threatening to a very young dairy calf. Viral scours tends to affect calves during the first two weeks of life, whereas bacterial scours can strike at any age up to about two or three months old. The youngest calves are usually most susceptible and most adversely affected.

*Symptoms/effects:* Scours is another word for diarrhea. The calf's manure is runny and watery.

*Prevention:* Calf scours is a complex problem. Diarrhea can be caused by many things, including bacteria, viruses, and protozoan parasites. Overfeeding a calf or using poor-quality milk replacers can also lead to scours. Poor nutrition and bad management that lead to a dirty, wet environment can make calves more susceptible to infections that cause scours. Get help from your vet to determine the cause of diarrhea, and how to treat it.

*Treatment:* Treatment for scours includes giving fluids containing electrolytes, and giving antibiotics in liquid or pill form.

A calf with diarrhea must be given fluids. In the early stages, while the calf is still strong, you can give electrolyte fluids with a nursing bottle if the calf will drink it. But a lot of calves will refuse to drink. You'll have to give the fluids by stomach tube or esophageal feeder. At this stage, oral fluids (by mouth, or into the stomach) are adequate and effective; the calf's gut is still able to absorb them.

## Early Treatment Can Save Your Calf

If you are feeding your calf twice a day, you will notice if it is ill. If you start treatment promptly, you should be able to halt the diarrhea in time to help the calf recover quickly.

*Electrolytes. Important body salts that are lost when a calf has diarrhea.*

But as the disease progresses, more gut lining is destroyed. The calf becomes weaker, unable to absorb fluid from the digestive tract, and more dehydrated. The only way you can save a calf this sick is with intravenous fluids given by your veterinarian.

Antibiotic pills for treating bacterial diarrhea are not as effective as a liquid antibiotic. A good liquid antibiotic for scours is neomycin sulfate solution. There are several different brands. One is called Biosul. This can be squirted into the back of the calf's mouth or added to the fluid-and-electrolyte mix.

If the antibiotic recommended by your vet is only available in pill form, you should crush or pre-dissolve the pills and give them as a liquid. Add enough water so the crushed pills can be given by syringe squirted into the back of the mouth, or added to your fluid mix.

Veterinarians used to say you should not feed milk to a scouring calf. But now vets have learned it isn't necessary to withhold milk. The calf should be encouraged to nurse its regular feedings, to keep up its fluid and energy. If the calf is too sick to nurse, or won't nurse enough, feed it the regular feedings by esophageal feeder. Give the medicated fluids and electrolytes in between.

For a dehydrated calf, the fluid-and-electrolyte mix should be given every six to eight hours, or until the calf feels well enough to nurse a bottle or nipple bucket again. The liquid antibiotic is needed only once a day, but make sure the calf is nursing or getting fluid three or four times a day to keep from getting dehydrated.

## Pneumonia

After scours, pneumonia is the most common killer of young calves. It can be caused by viruses or bacteria. The germs that cause pneumonia are usually lurking around. They only make the calf sick if its immunities are poor or if its resistance is lowered by stress. A

## Don't Mix Medications with Milk

Do not mix electrolytes with milk. Mixing medications with milk or milk replacer can prevent curd formation and can make the diarrhea worse. Wait two or three hours after the milk feeding before giving the fluid with electrolytes. Space the fluid treatments in between the regular feedings.

newborn calf that doesn't nurse soon enough or get enough colostrum (see page 114) will not receive the antibodies needed for adequate immunity. A young calf that has been weakened by a bad case of scours may come down with pneumonia. Young calves between the ages of two weeks and two months are very susceptible to pneumonia. Poor ventilation in sheds and barns can lead to pneumonia. Any severe stress can also make a calf of any age susceptible to pneumonia. Stressful conditions include overcrowding, wet and cold weather, sudden changes in weather from one extreme to another, a long truck haul, or bad weather during weaning.

***Symptoms/effects:*** Pneumonia can be swift and deadly, or it can be mild. If you can spot the warning signs early and start treatment quickly, pneumonia will be a lot easier to clear up. A calf coming down with pneumonia usually quits eating, lies around, or stands humped up looking depressed and dull. Its ears may droop. Respiration may be fast or labored and grunting. The calf may cough, or have a snotty or runny nose.

***Prevention and treatment:*** Confine the calf and take its temperature. If it's over 102.5°F, the calf is sick and should be doctored. A temperature over 104°F is very serious. Antibiotics should be given immediately, even if the illness is caused by a virus. Antibiotics will fight any secondary bacterial invaders; these are the deadly killers. Your vet should look at your calf and help you with the first doctoring. He may leave medication with you, along with instructions for treatment.

The calf will need good care. This means shelter to keep it warm and dry, and fluids, especially if it isn't eating or nursing enough. Fever can cause dehydration.

For a serious case of pneumonia, try to reduce the pain and fever. Your calf will feel better and start eating again. Dissolve two aspirin in a little warm water and squirt it into the back of the calf's mouth with a syringe.

Don't stop treatment too soon. The calf may be getting better, its fever may be down, and you might think you can stop giving it medicine. But if the symptoms return, the calf will be twice as hard to save. Give the antibiotics for at least two full days after all the symptoms are gone and the calf's temperature is

## Be Careful with Medications

When using any medications or vaccines, always read the labels and follow the directions for dosage, how to give it, etc. If there are any directions that you don't understand, ask your vet. Store medicines properly. Some need to be refrigerated; others should not be refrigerated. Some need to be kept out of the sunshine. Some need to be shaken well before use. Read the label!

*Immunity. The ability to resist a certain disease.*

normal again. It's not unusual to have to doctor a calf for a full week or longer.

Calves should be vaccinated against the most common virus-caused diseases such as IBR (infectious bovine rhinotracheitis), BVD (bovine virus diarrhea), and PI3 (para-influenza type 3). These often progress into pneumonia. The virus weakens the calf, and then bacteria move into the lungs and cause pneumonia.

## Blackleg and Other Clostridial Diseases

Blackleg is a serious disease caused by bacteria that live in the soil.

*Symptoms/effects:* Cattle get sick very suddenly, and most of them die.

*Prevention and treatment:* There is a good vaccine to prevent blackleg. Every calf should be vaccinated at about two months of age, with a second dose of vaccine around weaning time.

Blackleg is just one of several very serious diseases caused by a group of bacteria called "clostridia" that live in the soil. This family of diseases includes tetanus, "redwater," malignant edema, and enterotoxemia. There are vaccines that protect against several of these in the same injection. Check with your vet to see which vaccine you should use for your calf.

Some of these diseases are a problem only for calves. Once vaccinated, they have lifelong immunity. Others, such as redwater, can be deadly at any age. If redwater is a problem in your area (such as the Northwest), your calf must be vaccinated every six months.

## Brucellosis

Brucellosis is also called Bang's disease.

*Symptoms/effects:* This disease causes abortion in cows.

*Prevention and treatment:* All heifers must be vaccinated against brucellosis between 2 and 10 months of age. Steers do not need this shot. If your heifer was already vaccinated before you bought her, she will have a small metal tag in her ear with a number on it. The same number will be tattooed in her ear. If she doesn't have the tag and tattoo, ask your vet to vaccinate her.

## Leptospirosis

Leptospirosis is caused by bacteria and spread by mice, rats, and other rodents, as well as by wildlife or any infected domestic animals such as pigs, dogs, or other cattle. Cattle get "lepto" from contaminated feed or water.

*Symptoms/effects:* This is a mild disease in cattle, but it can have serious side effects, such as abortions in pregnant cows.

*Prevention and treatment:* You should vaccinate your calf against leptospirosis, especially if you are raising a heifer. Heifers and cows should be vaccinated at least once a year for lepto. Some veterinarians recommend every six months.

## Coccidiosis

Coccidiosis is a disease of calves caused by a protozoan — a tiny one-celled animal that damages the intestinal lining.

*Symptoms/effects:* Coccidiosis causes severe diarrhea. There is often blood in the loose, watery manure. The protozoans are passed in the manure of sick calves and "carrier" animals. Carriers are not sick, but have some protozoans living in their intestines. A calf may pick up the coccidia eating contaminated feed or water, or licking itself after lying on dirty ground or bedding.

*Prevention and treatment:* If your calf ever gets really loose manure — and especially if there is any sign of blood in it, or if the calf strains a lot after passing the loose bowel movements — have a vet examine it.

## Lump Jaw

Lump jaw occurs after a calf eats hay or grass with sharp seeds in it. Foxtail or downy brome ("cheat grass") seedpods have sharp stickers that can get caught in the

## Bang's in Wildlife

Requiring all heifers to be vaccinated against Bang's (brucellosis) has nearly eradicated this disease in cattle in the United States. However, Bang's is also spread by wildlife such as elk and bison. For instance, the bison in Yellowstone Park are infected with Bang's. So farmers and ranchers must continue vaccinating all their heifers.

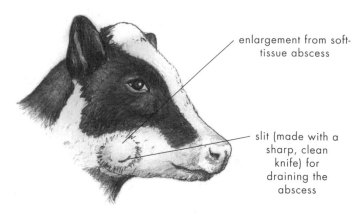

enlargement from soft-tissue abscess

slit (made with a sharp, clean knife) for draining the abscess

Lancing a cheek abscess

*Abscess. A pus-filled swelling.*

mouth, poking into the cheek tissue. If a sticker pokes in deeply, it may open the way for bacteria. Infection can develop. This may become an abscess.

**Symptoms/effects:** The abscess may get as large as a tennis ball, and you can see the calf's cheek bulging. That's why this disease is called lump jaw.

**Prevention and treatment:** The abscess must be punctured and drained. Sometimes abscesses break and drain on their own, but they heal faster and with less scar tissue if lanced and drained. Your vet or an experienced person should do this.

## Foot Rot

Foot rot is an infection caused by bacteria that live in the soil. Cattle may get the infection if they have a break in the skin of a foot. Wet areas, where cattle walk through mud, are the most likely places to get foot rot.

**Symptoms/effects:** Foot rot causes swelling, heat, and pain in the foot, resulting in severe lameness. The swelling and lameness come on very suddenly. Your calf may be fine at one feeding and very lame by the next. The foot may be too sore to walk on.

**Prevention and treatment:** Foot rot heals quickly if treated early with the proper antibiotics. If you know how to give injections, you can treat your calf for foot rot. Otherwise, get help from an experienced person.

Foot rot causes swelling between the toes.

## Diphtheria

Diphtheria is an infection in the throat and mouth, caused by the same bacteria that cause foot rot. Injury to tissues of the mouth or throat can let the bacteria in. Young cattle up through two years old are most susceptible. Emerging teeth (after the calf sheds its baby teeth) or injuries from coarse feed or sharp seeds can open the way for infection.

*Symptoms/effects:* The calf may be dull and not interested in eating. It may slobber and drool, with swellings in the cheek tissues. Its breath may smell bad. Swelling at the back of the throat can block the windpipe and make breathing difficult. The calf may have a cough, and will drool saliva because it has a hard time swallowing. The calf has a fever. It may die from the infection or from obstruction of its air passages unless it is treated quickly.

*Prevention and treatment:* Diphtheria can be very serious. If the symptoms are present, call your vet immediately.

## Pinkeye

Pinkeye is usually a summertime problem because the bacteria that cause it are spread by face flies. Pinkeye often occurs when dust, flies, sunlight, or tall grass irritate a calf's eyes. Flies carry the bacteria from one animal to another. Pinkeye may not be a problem if your calf lives alone, unless there are other cattle within flying distance for face flies. The disease is contagious.

*Symptoms/effects:* The calf holds the eye shut because it is very sensitive to light. The eye waters. After a day or two, a white spot appears on the cornea — the front of the eyeball. If it gets worse, the spot gets larger and the cornea becomes cloudy and blue.

*Prevention and treatment:* Mild cases clear up without treatment, but serious cases may cause blindness. So it's best to treat pinkeye and to make sure it heals up quickly.

The calf's tears wash medications out of the eye, so you must repeat the treatment at least twice a day. The eye can be treated with an antibiotic powder or spray. Your vet can recommend a good medication. Restrain or tie up the calf so you can squirt the medication right into its eye. Early detection and diligent treatment should clear up pinkeye within a few days.

## Bloat

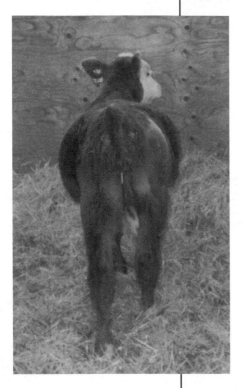

A bloated calf looks puffed up on its left side, where the rumen is located.

Bloat is a digestive problem that can be caused by highly fermentable feeds. Harmful bacteria that create gas when they multiply can also cause bloat. Too much gas builds up in the rumen. Burping may not get rid of the gas, especially if the gas is frothy (foamy or full of small bubbles). The tight rumen puts so much pressure on the lungs that the calf can't breathe.

*Symptoms/effects:* When viewed from behind, a calf looks very full or "puffed up" on its left side, where the rumen is located. As bloat gets worse, both sides puff up and the calf has trouble breathing.

*Prevention and treatment:* Several feed problems can lead to bloat, including lush alfalfa pasture or rich alfalfa hay. Too much grain, or finely chopped hay or grain, can also cause bloat.

If your calf bloats badly, the bloat must be relieved quickly. Your vet may pass a tube into the calf's stomach to let out the gas. If it is frothy bloat, which doesn't come out easily, the vet will pour mineral oil in through the tube, to break up the foam.

Some calves bloat often. If your calf has this problem you can feed Bloat Guard, which contains an antifoaming agent. It comes in block form, like a salt-mineral block. Give it in place of the salt block.

## Acidosis

A large increase in your calf's grain ration can result in acidosis — too much acid in the calf's body. Acidosis occurs when a sudden increase in grain causes over-production of lactic acid in the rumen.

**Symptoms/effects:** If not promptly treated, the rumen will quit working. The calf may get a fever, diarrhea, or founder (see below). It may die.

**Prevention and treatment:** Acidosis occurs most often when the grain ration is increased too rapidly. But it can also happen if something interferes with the calf's regular feeding schedule. This might happen if you forget to feed your calf and it gets too hungry and then overeats. If its water has manure in it, the calf may quit eating because it is thirsty from not drinking. Then after its water tub is cleaned and it drinks again, the calf may load up on feed. Fast action may be needed to save your calf or prevent founder. Call your vet immediately.

To prevent acidosis, have a schedule for feeding and stick to it. Split the daily grain ration and feed twice a day, so the calf isn't eating a large amount of grain at once. When increasing the grain ration, do it gradually, over a couple of weeks.

## Founder

A calf can founder if you feed too much grain or change its ration suddenly.

**Symptoms/effects:** The attachments between the hoof wall and sole of the foot become sore and may separate, resulting in malformed hoofs and severe lameness. The main cause is acidosis (see above).

**Prevention and treatment:** This is a serious emergency. Contact your vet immediately.

## Ringworm

Ringworm is a fungus infection, most common in winter. It is contagious, spread from calf to calf or from halters or brushes, or from calves rubbing the same posts.

**Symptoms/effects:** Hair falls out in 1- to 2-inch-wide circles, and the skin is crusty or scaly.

**Prevention and treatment:** Treatment should start immediately. The calf may have to be thoroughly washed with a solution that will kill the fungus. If there are only one or two scaly places, you can treat them with a fungicide recommended by your vet.

A calf with ringworm will not be allowed to enter a show. If you are planning to show your calf at the fair, try to clear it up quickly. Some types of ringworm are contagious to humans. Use rubber gloves and wash your hands with soap immediately after you treat your calf.

### Warts

Warts are skin growths caused by a virus. They usually affect calves and yearlings more than adult cattle, since mature cattle have usually developed some resistance to the virus. Warts are unsightly, but they will go away after a few months.

### Lice

Lice are active in winter, and can build up in large numbers on a calf.

*Symptoms/effects:* If your calf itches a lot, rubbing out hair over its neck and shoulders, it may have lice. When you are close you may be able to see the tiny black creatures around your calf's eyes or muzzle.

*Prevention and treatment:* Your vet, county Extension agent, or local farm supply store can recommend a product for getting rid of lice. Wear protective clothing, goggles, and rubber gloves when you apply it. Read the label before using the product. If it is a powder, do not breathe the dust. Apply it on a calm day with no breeze. You might want your parents or 4-H leader to help you delouse your calf. The powder will kill the lice, but not the louse eggs. Treatment must be repeated at least once to get the lice that have not hatched yet.

### Grubs

Cattle grubs, also called warbles, sometimes appear under the skin on the backs of cattle in late winter or early spring.

*Symptoms/effects:* Grubs look and feel like marbles under the hide. The grub is actually the maggot stage of the heel fly. This fly lays its eggs on the lower part of the legs of cattle during the warm days of early summer. The grub travels under the skin to the animal's back.

When heel flies bother cattle, the cattle may run wildly with their tails in the air. They may be so frantic they crash into fences. They also look for shade and try to stand in water holes to escape the heel flies.

*Prevention and treatment:* You may not have any heel flies, but if you ever do, consult your vet for the best way to get rid of them.

## Flies

Flies bite and suck blood, and also annoy and irritate cattle.

*Symptoms/effects:* If your calf spends all its time in the shade trying to get away from flies, it may not graze or eat as much as it would otherwise.

*Prevention and treatment:* Horn flies and face flies can often be controlled with insecticide ear tags. Insecticide is put into these plastic ear tags during manufacturing. While in the animal's ear, the tag continuously releases insecticide as it is rubbed against the hair. The calf rubs it over its body as it reaches around and scratches itself.

## Internal Parasites

"Worms" commonly infest cattle, especially young cattle that have not developed resistance to them. These parasites are most often a problem for cattle on pasture.

*Symptoms/effects:* The calf won't gain weight fast, or may even lose weight. It may have a rough hair coat, a poor appetite, diarrhea, or a cough. To tell if your calf has worms, have your vet examine a sample of the manure.

*Prevention and treatment:* Calves raised in a clean place will probably not get internal parasites. If your calf is alone in a small pen or hutch, and it was clean

*Parasite.* *An organism that lives in or on an animal.*

before it went in, you won't have to worry about having the calf dewormed. But once a calf goes out on pasture with other calves or where there have been other cattle, worms might become a problem. If your calf has worms, there is medication to get rid of them.

## Navel Hernia

Occasionally a calf will have a navel (umbilical) hernia. The abdominal wall at the navel area does not close up properly after the calf is born. There will be a bulge at the navel. Sometimes an enlargement there is caused by an abscess. If it is an abscess, it will be a firm swelling. A hernia will be soft tissue that goes back and forth through the hole in the abdominal wall.

A small hernia may go away by itself as the calf grows. But a large hernia is serious. A loop of intestine may come through it and strangulate. If this happens, the piece of intestine gets pinched off. The blood circulation to it is hindered, which can cause that portion of the intestine to die, killing the calf.

If your calf has a swelling at the navel, have a vet look at it. If it is an abscess, he can treat it by lancing, draining, and flushing with antibiotics. If it is a hernia, he can tell you whether it will get better on its own or whether he must correct it by putting in stitches to close up the hole so the tissue will grow together.

## Giving Injections

Many medications and most vaccines are given to cattle by injection with a syringe and needle. Most injections are given intramuscularly — deep into a big muscle. Some are given subcutaneously — under the skin, between the skin and the muscle — while others must be given intravenously — directly into a large vein.

Any I.V. (intravenous) medications should be given by your veterinarian. But you can learn how to give I.M. (intramuscular) or S.Q. (subcutaneous) injections. It's best if you have an experienced person show you how to fill your syringe, measure proper dosage, and give a shot.

## Restraining Your Calf

The calf should always be restrained before you give it a shot. Have the calf in a chute or tied up and pushed against the fence so it cannot jump around. If it is merely tied to a fence, it may kick you. Don't stand behind or beside the calf unless it is restrained so it cannot move.

## Intramuscular Injection

On calves, I.M. shots are given in the rump, the buttocks, or the thickest muscle of the neck. On larger animals, most injections should be given in the neck. Make sure the area where you will put the needle is very clean, without mud or manure — or the needle will take bacteria into the muscle with it. Wet skin and hair increase the risk of taking bacteria into the muscle with the needle.

### Safety Precautions

After you give an injection, discard the syringe and needle if they are disposable. If they are reusable, boil them before the next use.

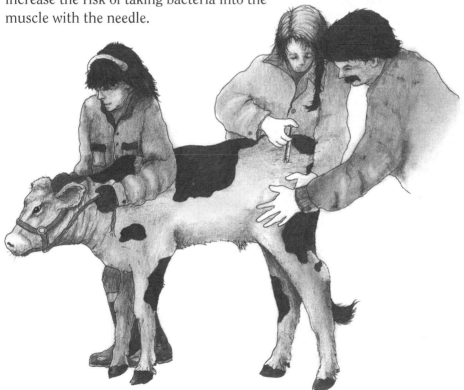

When giving your calf an injection, make sure the area you put the needle into is very clean.

Detach the needle from the syringe. Press the area firmly for a moment before putting in the needle. This tends to desensitize the skin; the calf will not jump so much when you jab it. Put the needle in with a forceful thrust so it will go through the skin easily.

A new, sharp needle goes in with less effort, and causes much less pain for the calf. If the calf does jump, you can wait until it settles down again before you attach the syringe and give the injection. Also, you can tell whether or not you've hit a vein. If the needle starts to ooze blood, take it out and try again in a different spot. Don't ever inject into a vein.

## Subcutaneous Injection

For a subcutaneous injection, lift a fold of skin on the shoulder or neck, where the skin is loosest, and slip the needle in. Aim it alongside the calf so that it goes under the skin you have pulled up, but not into the muscle. With a little practice you'll find that S.Q. shots are very easy to give.

## Giving Oral Medications

Have your calf tied or in a chute to give it pills or liquid medication by mouth, so it can't swing its head away or hit you with its head while trying to avoid the medicine.

## Pills

Pills can be given with a balling gun. This is a long-handled tool that holds the pill while you put it toward the back of the calf's mouth. Then you press the plunger and it pushes the pill out of its slot into the mouth. If you put the gun far back in your calf's mouth, the calf must swallow the pill when the balling gun releases it. This tool keeps your fingers from being bitten.

## Liquids

Giving liquid medication by mouth is easy using a big syringe without the needle. Fill it to the proper dosage

A balling gun is a long-handled tool that makes it easier to give your calf a pill.

and slowly squirt the medication into the back of the calf's mouth. Stick the syringe into the corner of the mouth and aim it far back.

You can give liquid antibiotics or pills dissolved in water this way. If the dose is large, give it a little at a time, allowing the calf time to swallow each portion before you squirt in the rest. Keep the calf's head tipped up so the medication cannot run back out.

## Proper Care Prevents Illness

By giving good care and paying close attention to your calf, you can help make sure it stays healthy, and you'll be able to detect any signs of illness early.

## Keep Your Calf Comfortable

To prevent many types of illness, avoid stress and feed your calf properly. Make sure it has shelter from cold weather and wind. It also needs shade during hot weather. A severe heat wave in summer may kill it if temperature and humidity get too high.

### Don't Get Bitten!

Be careful when examining the inside of your calf's mouth or giving pills. A calf has no top teeth in front, but can still hurt your fingers between the molars (back teeth) if your calf bites down.

## Keeping Cool

During very hot weather you can install a fan in the stall, or hose your calf down periodically with a misty wet spray from your garden hose.

## Sanitation

Good sanitation is always important. It is easier to try to prevent infectious diseases than to cure them.

If there are several calves in a pen or barn, or if there have been other calves there before, bacteria and viruses are always lurking around. You can help prevent diseases and infections by thoroughly cleaning and disinfecting your facilities between calves or between groups of calves. Get rid of all the old bedding, and scrub the walls (and the floor, if there is one) of the shelter with a good disinfectant. Your vet or a dairyman can recommend one.

# PART TWO

# Raising a Beef Calf

# CHAPTER 5

# Choosing a Beef Breed

**B**eef breeds in this country are descendants of cattle imported from the British Isles (Scotland, England, and Ireland), or from European countries (referred to as "continental breeds"), or from India. Many modern breeds are mixes of these imported breeds. The first cattle came here from Spain, but they were soon outnumbered by British cattle.

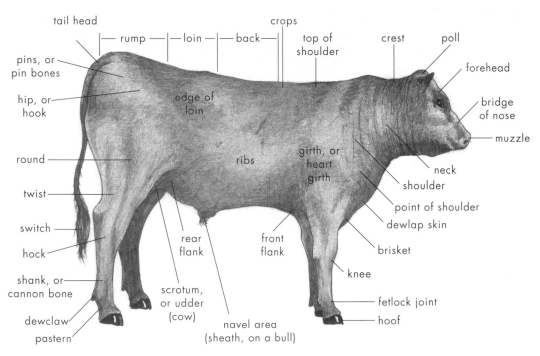

Parts of a beef animal

## British Breeds

British breeds are the ones that originated in the British Isles — England, Scotland, and Ireland.

### Hereford

The Hereford is well known for its red body with white face. The feet, belly, flanks, crest (top of neck), and tail switch are white. Other characteristics of the Hereford are large frame and good "bone" (heavier bones than many breeds). It has a mellow disposition, compared with Angus or some of the European breeds.

The Polled Hereford is identical to the Hereford except that it has no horns.

### Angus

Angus cattle are black or red and genetically "polled" (always born hornless). Smaller and finer-boned than Herefords, Angus are known for calving ease — they give birth to small calves. This characteristic makes them popular for crossing with larger, heavy-muscled cattle. They are also noted for early maturity, marbling of meat, and motherliness. Angus cows are aggressive in protecting their calves and give more milk than Herefords.

### Shorthorn

Shorthorn cows have good udders and give a lot of milk. They have few calving problems. Even though

Hereford

Polled Hereford

Angus

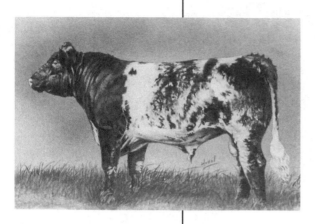

Shorthorn

**Roan**. *White hairs mixed with hairs of another color.*

## Cattle in North America

Christopher Columbus brought cattle to North America. The Vikings may have brought some even earlier, but those were probably all butchered for food. Spanish explorers and settlers brought long-horn cattle. Later the Pilgrims and other New England settlers brought cattle from England.

calves are born small, they grow big very quickly. Shorthorn cattle can be red, roan, white, or red-and-white spotted.

## Galloway

Galloways are hardy, with a heavy winter hair coat. Cows live a long time, often producing calves until age 15 or 20. The calves are born easily because they are small, but they grow fast. Most Galloways are black, but some are red, brown, white (with black ears and muzzle and black feet and teats), or belted (black with white midsection). Galloways are polled.

## Scotch Highland

Scotch Highland cattle are small, with long shaggy hair and impressive horns. These hardy cattle do well in cold weather and their shaggy coats also provide protection from insects; their long forelocks protect their eyes from flies.

## Dexter

Dexter are probably the smallest cattle in the world, and are used for milk and beef. The average cow weighs less than 750 pounds and is only 36 to 42 inches high at the shoulder. Bulls weigh less than 1,000 pounds and are 38 to 44 inches tall. They are quiet and easy to handle, and the cows give very rich milk.

## Continental Breeds

There are many European beef breeds in America. They have become popular in the past 30 years as exotics, adding size and muscle (and sometimes more milk) to crossbred herds in this country.

## Charolais

Charolais are white, thick-muscled cattle. Cattlemen in this country like Charolais for crossbreeding because they are larger than British breeds.

## Simmental

Simmentals are yellow-brown cattle with white markings. They are known for rapid growth and milk production, good for making butter and cheese.

## Limousin

The Limousin is a red, well-muscled breed. Cattlemen like the moderate size and abundance of lean muscle, as well as calving ease. The small calves are born easily and grow very fast.

## Tarentaise

Tarentaise are a breed of cherry red cattle with dark ears, nose, and feet. They are moderate-sized dual-purpose animals, used for milk and meat, with early maturity and good fertility.

## Salers

Salers cattle are horned, dark red in color, and popular in America for crossbreeding, due to good milking ability, fertility, calving ease, and hardiness.

Charolais

Simmental

Limousin

Salers

Gelbvieh

Texas Longhorn

### Chianina

Chianina are the largest cattle in the world. They are white in color.

### Gelbvieh

Gelbvieh are light tan to golden. Gelbvieh calves grow fast, and the heifers mature more quickly than most of the other continental breeds.

### Other Continental Breeds

There are many other continental breeds in the United States today, including Maine Anjou, Pinzgauer, Piedmontese, Braunvieh, Normandy, and Romagnola. In general, the continental cattle are larger, leaner, and slower maturing than the British breeds.

### Breeds from Other Places

Some of the cattle breeds in the United States originated in places other than the British Isles and the European continent.

### Murray Grey

The Murray Grey is a silver-gray breed from Australia. They are gaining popularity in America because of their moderate size, gentle dispositions, and fast-growing calves. The calves are small at birth but often grow to 700 pounds by weaning.

### Texas Longhorn

The Texas Longhorn is descended from wild cattle left by early Spanish settlers in the Southwest. Longhorns are moderate sized, and well known for calving ease, hardiness, long life, and fertility.

### Brahman

Brahman cattle are easily recognized by the large hump over the neck and shoulders, loose floppy skin on the dewlap and under the belly, large droopy ears, and horns that curve up and back. They come in a variety of colors.

Brahmans do well in the southern part of North America because they can stand the heat and are resistant to ticks and other hot-climate insects. They are large cattle, but calves are very small at birth, growing rapidly because the cows give lots of rich milk.

Brahman

### Crossbreds

There are nearly 100 cattle breeds in the United States. Cattlemen often cross them to create their own unique cow herds with the traits they want. A crossbred is an animal with parents of different breeds. Crossbreeding is a very useful tool for the beef producer.

### Breeding for Different Environments

Cattle are raised in a wide variety of environments — from lush green pastures to dry deserts and steep mountains. Each farmer or rancher tries to create a type of cow that will do well and raise good calves in his situation. No single breed is best in all the traits that are important to beef production.

### Grazing Cattle

Beef cattle can graze on land that won't grow crops. More than 90 percent of the 810 million acres in America used for cow pasture are too rough and steep, too dry, too wet, or too high to grow food crops. So beef cattle are a good way to use these lands to create food.

## Hybrid Vigor

The most effective genetic advantage in cattle breeding is hybrid vigor, which results when you mate two animals that are very different. This is only obtained through crossbreeding. Hybrid vigor increases fertility, milk, and life span of cows, vigor and health of young calves, and many other important traits. With careful crossbreeding, the rancher can develop crossbred cows that will do better than the parent breeds. Good crossbred females make the very best beef cows.

## Composites

"Composite" cattle are blends of different breeds into a uniform type of crossbred. There are several composites that have been created in the last 20 years and more new ones are being formed all the time. Nearly every breed we have today began as a composite.

Brangus, Santa Gertrudis, and Beefmaster are examples of successful composites. More recent blends are the Hays Converter (a Canadian breed made up of several beef breeds combined with Holstein — a dairy breed), the RX3 (a blend of Hereford, Red Angus, and Holstein), and numerous others.

## How Shall I Choose?

There are so many breeds and crosses to choose from! Before selecting a calf of a certain breed or combination of breeds, you should consider several things.

### Evaluate Strong Points and Faults

All breeds have strong points and faults. No one breed is perfect for all situations. This is why crossbreds and composites are becoming more popular. There can be vast differences among individual animals within a breed, so examine each calf carefully.

## Ask for Help

Whether you are looking for a heifer or a market beef steer, there are people who will be very willing to help you find one. Your county Extension agent or 4-H leader can tell you which breeds and crossbreds are being raised in your area and help you contact the stockmen who are raising the kind you'd like.

## Evaluate Local Conditions

When selecting a calf, try to make an objective decision about breeds. Evaluate the region where you live, to see which breeds are best suited for local conditions. A woolly breed might not do well in a hot climate, and a Brahman wouldn't do well in the cold.

## Registered Purebreds or Commercial Cattle?

If you are going to raise a heifer to start a breeding herd, decide whether to have registered purebreds or commercial cattle. Commercial cattle can take advantage of the benefits from crossbreeding. If you want to have a purebred heifer, it's often best to select a breed being raised in your area, so you can use bulls from someone's herd close to home. Most purebred breeders are glad to help young people get started. If you buy a heifer from a registered breeder, he will probably help you get her bred.

**CHAPTER**

# Selecting a Beef Calf and Bringing Him Home

## Purebred

The word "purebred" refers to an individual of a breed (such as a purebred Angus) that has no other breeds in his ancestry. A purebred is not necessarily registered. The term purebred should not be confused with Thoroughbred, a certain breed of horse.

A registered purebred has a registration number, recorded in the "herd book" of a breed association. The association gives the owner a certificate stating that the animal is the offspring of certain registered parents.

**Y**ou've learned about the basic care of calves. You've prepared a place to keep a calf of your own. Now it's time to choose your beef calf and bring him home.

### Where Can I Get a Calf?

You can buy your calf at a local auction, at a feeder calf sale in the fall, at a special "club calf" sale, or at a farm or ranch. A local purebred breeder or commercial cattle producer would probably be a good source.

One advantage to buying directly from the person who raised the calf is that it is easier to find out things you need to know, such as what vaccinations the calf already had and when they were given.

### Find Out About Prices

Before you go to an auction sale or visit a ranch, find out what beef calves are selling for that week, per pound. You don't want to pay more than the market price, especially if you are hoping to make a profit when you sell him as a yearling.

Some 4-H kids pay high prices for a calf, hoping to get a future champion, but this is always a risk. A good calf should bring top market price, but no calf is actually worth more than that, even if he is a potential champion. And if he's not a champion and doesn't bring a high price at the 4-H stock show sale, you would lose money on your project.

## Picking Out Your Calf

A beef steer doesn't need to be purebred. The best beef steers are usually crossbred, a mix of two or more breeds. A good crossbred steer often grows faster, with better feed conversion (more pounds of beef produced, on less feed) than most purebreds.

The best steer for a beef project is a fast-growing, well-muscled animal that will reach a market weight of 1,050 to 1,250 pounds by the time he is 14 to 20 months of age. If you are picking a steer for a 4-H project, you don't want him too small or too large at his finish weight. Steers of the larger breeds may weigh more than 1,300 pounds by the time they finish. Choose a steer that will finish within the weight requirement for your county fair. If this is your first experience, have a 4-H leader help you pick out the proper kind of calf.

## Frame Score

Beef cattle are categorized by "frame score," which is a way of saying whether they are small-, medium-, or large-bodied. A small-framed early-maturing steer will not produce enough meat on his small carcass. If you try to get him to grow bigger, he'll just get too fat; he is not genetically capable of growing bigger. A very large-framed steer will grow too big before he gets fat enough to butcher, taking too much feed. The most practical kind of beef steer has a medium frame.

The calf you pick should have a lot of muscle, not a lot of fat. He should have nice smooth lines, and not be sway backed. He should have a deep body, not shallow,

## Not Too Thin, Not Too Fat

Thinness could be a clue that he has been sick or is not healthy. And don't pick a really fat one. If he is already fat, he may not grow as well; he may finish out too quickly.

*Finish weight. The weight at which a beef steer has enough fat and is ready to be butchered.*

but not pot bellied either. He should be long and tall, but not extremely tall. Your beef steer or heifer must be large enough at the start of your project to reach the desired finish weight by fair time.

## Figuring Frame Score for Beef Calves

To figure your calf's frame score, measure his height from the ground at the hip when he is standing squarely. Then look up his age on this chart and find the hip height on that age line. Look to the top of the column for the frame score. For instance, a 10-month-old calf that is 45 inches tall at the hips would be a frame score 4.

| Age (months) | Frame Score (hip height in inches) | | | | | | |
|---|---|---|---|---|---|---|---|
| | 1 | 2 | 3 | 4 | 5 | 6 | 7 |
| 5 | 34 | 36 | 38 | 40 | 42 | 44 | 46 |
| 6 | 35 | 37 | 39 | 41 | 43 | 45 | 47 |
| 7 | 36 | 38 | 40 | 42 | 44 | 46 | 48 |
| 8 | 37 | 39 | 41 | 43 | 45 | 47 | 49 |
| 9 | 38 | 40 | 42 | 44 | 46 | 48 | 50 |
| 10 | 39 | 41 | 43 | 45 | 47 | 49 | 51 |
| 11 | 40 | 42 | 44 | 46 | 48 | 50 | 52 |
| 12 | 41 | 43 | 45 | 47 | 49 | 51 | 53 |
| 13 | 41.5 | 43.5 | 45.5 | 47.5 | 49.5 | 51.5 | 53.5 |
| 14 | 42 | 44 | 46 | 48 | 50 | 52 | 54 |
| 15 | 42.5 | 44.5 | 46.5 | 48.5 | 50.5 | 52.5 | 54.5 |
| 16 | 43 | 45 | 47 | 49 | 51 | 53 | 55 |
| 17 | 43.5 | 45.5 | 47.5 | 49.5 | 51.5 | 53.5 | 55.5 |
| 18 | 44 | 46 | 48 | 50 | 52 | 54 | 56 |

*Conformation. The general structure and shape of an animal.*

## Disposition

The disposition, or personality, of your calf is just as important as his weight and conformation. Some calves are more placid and easygoing than others. If you are getting only one calf, try to select a mellow individual, not a nervous one that will "go bonkers" being by himself. Choose a smart and gentle one that will learn to trust you and become your friend.

## Making a Place for Your Calf

Before you buy your calf, prepare the place where you'll keep him.

### A Strong Pen

If your calf will be living by himself, have a strong pen to put him in for a few days before you turn him out to pasture. Be sure the fence is something he cannot jump over or crawl through. A frantic, homesick calf in a new place may try to jump over or crash through even a very good corral fence, so make sure your pen is calf-proof.

If the calf you purchase has already been weaned, he won't be so desperate to try to crawl out and go back to his mother.

### Check for Hazards

Make sure your pen or pasture has no hazards such as nails or loose wires that might injure your calf. A pole or board on the ground with nails sticking out of it could cause serious injury if the calf steps on it. Don't leave any baling twines hanging on a fence or lying on the ground, because the calf may chew on them.

Calves are curious, just like little kids, and often get into trouble. If your calf tries to eat baling twine or to chew on a plastic bag that blew out of your garbage barrel, pieces may plug his digestive tract and kill him. He may also chew on electrical wires in the barn. Make sure there is nothing within his reach that might cause him harm. Wire or nails lying around near his feed may puncture his stomach if he eats them, causing "hardware" disease (often fatal).

### Don't Choose a Wild One

A wild, snorty calf is a poor risk, even if he is big and beautiful. He is difficult to halter break, tie up, and lead, and you may have trouble trying to gentle him. He could also be dangerous, knocking you down or kicking you. A wild one won't gain weight as well as a placid calf. Rate of gain (pounds gained per day) is almost always better with a gentle calf.

## Getting Your Calf Home

Make arrangements with someone who has a trailer or a pickup truck with a rack to haul your calf home. If you are buying the calf from a farmer or rancher, he may be able to haul the calf for you. Ask what he would charge.

If you will be unloading the calf into a pen or pasture, a trailer often works best, because it is low to the ground and the calf can step out easily. A truck is too high; the calf must be unloaded at a loading chute. Even a pickup with a rack is often too high for a calf to jump out of without risk of injury, unless it can be backed up to a bank, or you have a ramp.

When buying a beef calf, remember that most of them have lived with their mothers in large pastures. Some may not have seen any people close up, except for vaccinations or medical treatment. Their experiences with people have probably been scary and painful.

Your calf may even try to run over you if you get in his way as he comes out of the truck or trailer. Have an experienced person bring your calf home and help you unload him.

If this is not possible, your family may have to borrow a trailer to transport the calf. If you buy the calf at an auction and load him into a stock trailer, the sellers can help herd the calf into the trailer from the loading alley. Likewise, the farmer will help you load a calf at his place.

## Safety First

When working with cattle, make a habit of thinking "safety first." Before you do anything, ask yourself what the calf might do. He is a large, strong animal, and he will need time to settle down and lose his fear, and get to know you. After he knows you, it will be a lot easier for both of you. So at first, take your time, play it safe, and give him time to get adjusted to his new place, and to you.

## Unloading

When you get home, make sure the trailer is backed up into the pen far enough so the calf cannot go anywhere except into the pen. The pen gate should be swung tight against the trailer. A scared calf may try to bolt through even a small opening. Don't stand in a place where he might run over you. If you are unloading your calf from a pickup, make a ramp from sturdy boards.

# Caring for Your New Beef Calf

# CHAPTER 7

**Y**ou've unloaded your calf and he's now in your pen — his new home. He's probably lonely and scared, unless he has another calf for a buddy. Try not to frighten him, or he may become so upset and frantic he might crash into the fence.

## Understanding Your Calf's Behavior

To understand your new calf, try to think like a calf. Cattle are herd animals. They are happiest when they can be in a family group with other cattle. Your calf has probably not spent much time with people.

## How Your New Calf Feels

If your calf was already weaned before you bought him, he's already gone through the emotional panic of losing his mother. He will miss the other calves he was with, but he won't be quite as desperate to get out of your pen to go find his mother. He'll just need time to adjust to his new home.

But if your calf is still going through weaning when you bring him home, it will take him several days of very stressful adjustment. He may pace the fence and bawl. He may show little interest in feed or water.

## Safety in Numbers

Wild cattle were safer in herds. If wolves approached, the cows would bellow and all come running to form a tight group.

That's why yearlings and young cattle generally travel in a group. If one goes to water, they all do. If the leader decides it's time to go graze, they all go. It's not that they are just copycats; they are doing this for protection.

A calf being weaned is more susceptible to illness, because stress hinders the immune system. If the weather is cold, rainy, or windy, a weaning calf may be particularly prone to pneumonia.

## Getting Acquainted

When you are starting to get acquainted with your new calf, give him time and space. Don't try to get too close to him. Until he gets to know you, he may react explosively if he feels cornered.

Speak softly. Move slowly. When you approach his pen to feed him, let him know you are there. If his attention is diverted elsewhere and then he suddenly sees you, he may run off. Talking softly or humming a little tune can help to gentle a wild calf.

When you are in the pen with your calf, don't look directly at him. He'll relax more if you act like you aren't paying attention to him. If you come too close, too fast, and look directly at him, he will think of you as a predator. Instead, ignore him but talk softly as you go by. Pretty soon he'll come eagerly to meet you when you bring his feed.

## Feeding Your Calf

When you first bring your calf home, have feed and water for him in the pen. Leave some good hay where he can find it easily, but not in a corner or along the fence line where he'll walk on it every time he goes around his corral looking for a way out. Use really good grass hay, or a mixed grass-and-alfalfa. Don't give him rich alfalfa hay at first, because it may make him sick or bloated. You can gradually adjust him to good alfalfa hay later.

### Water

Give him water in a tub or bucket hooked to the fence or in a water trough so he can't tip it over. If he's never drunk from a tub or bucket, you may have to put it next to his feed, or feed him next to his water trough for the first day or two, so he'll find the water when he comes to eat the hay. If he grew up in a pasture where his only experience with water was in a stream, pond, or ditch, he might not know how to drink from a water tub.

## Vaccinations

Check with the former owner to find out what vaccinations your calf has had. He will need several, to give him immunities against certain diseases. If the calf you buy has not had these injections, he will need them soon. He may also need a booster shot if his vaccinations were given when he was just a small baby. Ask your 4-H leader, county Extension agent, or veterinarian what vaccinations are needed in your area, and have someone help you vaccinate the calf.

## Handling and Gentling Your Calf

Some calves are not as timid as others and will be curious about you from the beginning. Use this curiosity to your advantage. If you are patient and quiet, the calf will come closer to you.

### The "Flight Zone"

Cattle have a certain space in which they feel secure. This is their "flight zone." It's like a big imaginary bubble. As long as you don't enter this zone, they feel safe. But if you get too close, they get nervous or scared and run off.

Different cattle have different-sized zones. A wild or timid calf has a large one; a gentle or curious calf has a much smaller one. Once your calf gets to know you well, his flight zone will disappear.

### Use Feeding to Your Advantage

When he begins to associate you with food, your calf will lose his fear and come right up to you. It may still

## Making Friends

Cattle like to be petted and scratched. They especially enjoy it in places that are hard for them to reach. Most love to be scratched under the chin, behind the ears, and at the base of the tail. But don't rub the top of the head or the front of the face. This will only encourage your calf to bunt at you.

take a few more days before he'll let you touch him, but he will stand there beside you and eat.

Cattle are good at associating things. If you have a special call for feeding time, he'll come to you every time he hears it. If you wish, you can give him a name, and use it when you call him. He'll come whenever he hears you call his name, for he'll expect something good — either food or a nice chin scratching.

Andy Rida adjusts the halter on his calf.

## The Importance of Trust

Don't upset your new calf by having all your friends come to see him right away. Wait until he has adjusted and gentled down a bit. Once he trusts people, it's fine to show him to your friends. Some calves figure things out quickly and soon love to be handled by people, while a nervous calf will take a bit longer.

## Don't Spoil Your Calf

Don't make the mistake of babying your calf too much. He should trust you, but he must also respect you. Remember that cattle are social animals and accustomed to life in a group — bossing other cattle or being bossed. He thinks of you as one of the herd. You must be the dominant herd member; he must accept you as the "boss cow." Otherwise he will try to be too bossy and pushy.

If your calf ever starts pushing on you or bunting at you with his head when you are feeding or petting him, discipline him with a swat. This is their version of play. Calves fight each other when they play. So your calf will naturally want to "play fight" with you.

If you spoil your calf by letting him do whatever he pleases, you will regret it later. If he gets too sassy, you can carry a small stick when you feed him or work with him, and rap his nose when he misbehaves. This will remind him you are the boss.

## Be Careful

Though a calf is not likely to attack a person, he can accidentally hurt you because of his size and weight. Always keep an escape route for yourself in mind when trying to corner or work with your calf — enough room to dodge aside if he backs into you or turns around and runs back out of a corner.

Cattle can be dangerous when handled in a confined area, because they may panic. Don't wave your arms, scream, or use a whip. If your calf won't move forward into the catch area, prod him with a blunt stick, if necessary, or twist his tail to get him to move. Just be careful to not twist it too hard. You can twist it into a loop, or push it up to form a sideways S curve. If you have to twist his tail to get him to move, stand to one side so he can't kick you.

## Be Gentle

Don't yell or chase your calf or you may scare him badly. Even if he is stubborn or suspicious and won't go into the catch pen or behind the gate or panel on the first try, don't get impatient with him. If you lose your temper and start yelling, this usually just confuses or scares him and makes things worse. He'll be harder to handle next time.

# Feeding Your Beef Calf

**Y**ou'll feed your calf on hay and pasture if you are raising a beef steer inexpensively or if you are raising a heifer to be a cow. You'll feed your calf grain and hay if he is a 4-H market steer or show steer.

If you want to show your steer, you should join a 4-H club that has beef projects. The club leaders and older members can help you with your project.

## Weighing Your Calf

You need to weigh your calf at the start of your 4-H project. The club leader will help you with this, making arrangements for transportation of calves to be weighed at a livestock scale. You need to know what your calf weighs at the start and finish of your project. That's how you figure out how much he gained, and his rate of gain.

## Basic Nutritional Requirements

Your calf will need water, grain, and roughages (hay or pasture). He can grow up on roughages alone, but will grow faster and get fatter sooner if you feed him grain. Grain is called a "concentrate" because it has more calories per pound than grass or hay. (See chapter 2 for more information about nutrition.)

## Expected "Finish" Weight for a Steer

| Breed | Angus | Hereford | Shorthorn | Charolais | Simmental | Limousin | Holstein |
|---|---|---|---|---|---|---|---|
| Angus | 1,000 | 1,050 | 1,050 | 1,225 | 1,225 | 1,125 | 1,225 |
| Hereford | | 1,050 | 1,075 | 1,250 | 1,250 | 1,150 | 1,250 |
| Shorthorn | | | 1,050 | 1,250 | 1,250 | 1,150 | 1,250 |
| Charolais | | | | 1,400 | 1,425 | 1,325 | 1,425 |
| Simmental | | | | | 1,400 | 1,325 | 1,425 |
| Limousin | | | | | | 1,200 | 1,325 |
| Holstein | | | | | | | 1,400 |

*Heifers of the same breeds and breed crosses would weigh about 80 percent of these figures.*
*If your calf is a cross of two of these breeds, look at the figure where their charts meet. For instance, the top line shows the weights of Angus and Angus crosses.*

## Hay and Grain for Your New Calf

When you start feeding your newly purchased calf, give him all the hay he will eat. Then slowly start him on grain, giving just a little bit until he learns to eat it, and then increasing the amount gradually. Too much grain all at once may upset his digestion.

## Clean Water

Make sure your calf has a constant supply of clean water. He won't eat enough feed if he gets thirsty. He will do better and gain faster if he always has plenty of water.

If he has a ditch or pond to drink from, you need to check it regularly, especially in winter when you may have to break ice. If there's too much ice, it's better to water him in a tub.

If he drinks in a ditch during summer, make sure it is always running. If he depends on you for water, make a habit of filling his tub or tank as often as necessary. If you have an automatic waterer, be sure it is working properly.

The bigger your calf, the more water he needs. All cattle need more water during hot weather than in cold weather.

Make sure your calf always has a source of clean water to drink.

## Feeding a Show Steer

If you are preparing for a 4-H show or market beef class, the calf will need high-quality feeds to gain market weight in a short time. In 4-H you are feeding a calf for a particular marketing date or show, trying to have him at the correct weight and finish by that date.

## Figuring a Calf's Rate of Gain

A market steer should weigh 1,000 to 1,300 pounds, with about 0.25 to 0.45 inch of outside fat on the carcass, with the grade "Choice" (see box at right) when sold. A market beef heifer will finish at a lighter weight (about 900 to 1,000 pounds) than a steer of the same age.

Most beef animals eat about 7 pounds of feed to gain 1 pound of weight. To some extent, a calf's rate of gain depends on his genetics, and some cattle have better rates of gain than others. On average, a growing steer should be able to gain 2 to 3 pounds per day. Some crossbred steers will gain more. You will need to feed your calf 15 to 20 pounds of feed per day.

By knowing the desired weight for your calf at the end of your project, and what he weighs at the beginning, you can calculate the gain needed to get him ready for the show or sale. Knowing the total number of days until that show or sale, you can determine the necessary average daily gain (ADG) needed.

For example, if you bought a 500-pound steer in November, with the 4-H show date 270 days away in August, you can use the following formula to figure out how much your calf must gain per day to finish at 1,100 pounds: Market weight, minus present weight, divided by number of days until show or sale, equals his ADG. For example, 1,100 pounds, minus 500 pounds, divided by 270 days, equals 2.2 pounds per day for his necessary daily gain.

Good average daily gain for a steer is 2.5 to 3.0 pounds per day. The steer in the example should have no trouble meeting his finish weight. Good ADG for a young heifer is 1.5 to 2.0 pounds per day.

## Grades of Beef

Beef cattle carcasses are inspected by the USDA (U.S. Department of Agriculture) and judged for quality, using several grades to rate the tenderness of the meat. The main thing that determines the grade is the amount of marbling — flecks of fat in the muscle — which makes the meat more tender, tasty, and juicy. The highest grade is "Prime," followed by "Choice," "Select," and "Standard." Prime has the most marbling; Standard has very little.

**Genetics.** *Qualities and physical characteristics that are inherited from parents and ancestors.*

## Cold-Weather Feeding

In cold weather, your calf will need more feed to generate body heat, to keep himself warm. Roughages provide more heat (from the fermentation process of digestion) than do grains. If the weather is cold, increase his ration of grass hay.

## Feeding a Beef Heifer

If your beef project is a heifer instead of a steer, and you are going to show her at the fair and market her as beef, you will feed her like you would a steer. Just take into consideration that she will not finish out as large as a steer of her same age and breed, and therefore will not need as much feed. Adjust your feeding figures to fit her target finish weight.

## Don't Feed Your Heifer Too Much Grain

If you are raising your calf as a breeding heifer project or planning to keep her as a cow, you should not feed her much grain. You want her to grow well, but not become fat. If she has too much fat deposited in her udder, she will never milk very well. Fat displaces the developing mammary tissue (the milk-producing glands). Too much fat around her internal organs will make her less fertile. She may not breed as quickly as a leaner heifer, and might not become pregnant even when bred. If she is too fat during pregnancy, she may have trouble calving. It will be harder for her to push the calf out. A cow will live longer and stay healthier if she is well fed but never overfed.

Some people feed heifers grain to have them gain weight faster, especially in purebred herds. But most commercial cattlemen, who depend on what their cows will produce without pampering, raise crossbred cattle that grow fast on just hay and grass. If a heifer must have grain or expensive protein supplements to get big enough quickly, this takes the profit away from the calf she will produce.

## Feed Efficiency

If you are raising a heifer to start a cow herd, choose one that can grow nicely and breed quickly without grain. Most good crossbred heifers will do this, and also some purebreds.

A clue to the genetic ability of your heifer for feed efficiency can be obtained by seeing what type of feeding management is used in the herd she came from. Ask the rancher how he raises his heifers — whether he grows them on grass and hay or feeds them grain.

## Rate of Growth for a Heifer

Your heifer calf needs good feed to grow big enough to reach puberty at an early age and be ready to breed by the time she is 15 months old. She needs to have 65 percent of her mature size by 14 months of age.

The actual desired weight at this age will depend on your calf's breed. Angus cattle mature at a lighter weight and are ready to breed at an earlier age than Herefords or Simmentals, for instance. Know what your heifer should weigh for her age and breed and feed her accordingly.

If you are raising your heifer as a 4-H project, you need to keep track of her rate of gain and have a target weight — a goal to shoot for — at various phases of the project. For example, one goal might be an ideal breeding weight.

To be the proper size and maturity for breeding, British-breed heifers should weigh at least 650 to 700 pounds, and European breeds should weigh as much as 800 to 850 pounds. Weigh your heifer or use a weight tape (see pages 154–155) and figure out how many days are left before you want to breed her. Then you can find out how much she has to gain to meet that weight. Finally, figure what her average daily gain should be.

*Puberty.* *The age at which an animal matures sexually and can reproduce.*

### Estimating a Heifer's Weight

An easy way to estimate her weight is with a weight tape. You measure around her girth area and it tells you approximately how much she weighs.

# Training, Fitting, and Showing Your Beef Calf

**E**ven if you don't plan to show your calf, teach him to tie up and lead. It will make your calf easy to handle in all the things you will be doing with him.

Fitting and showing a calf can be fun and rewarding. Going to shows gives you the opportunity to meet people with your same interests, make new friends, exchange ideas, and learn a lot.

## Halter Breaking Your Calf

If your calf is large and not very gentle when you first bring him home, you may need help handling him at first, from parents or someone with experience handling calves. Let him have a few days to adjust to his new place and get acquainted; then start working with him.

### A Small Place

You need a small place where you can get close to your calf — a small pen to lock him in, or a shed or stall.

*Fitting.* Clipping, washing, and brushing an animal for show.

## Equipment for Training

Equipment you'll need for training your calf: A nylon rope halter, a brush or comb, a show stick. The show stick is used for tapping your calf to give him signals, to rub him with, and to move his feet.

Herd your calf gently into the stall, shed, or small pen. Get him accustomed to being locked in there while he eats. If he's nervous, let him eat in the small area for a few days before you try to halter him. That way, he'll get used to confinement and begin to feel at home in there.

If you have several calves in a group, an easy way to gentle them is to herd them all into a small stall where they are crowded together. Enter the stall slowly. Gently scratch the calves on the back, with your hand or a scotch comb. Don't touch their heads, which would alarm them more. At first they will try to get away from you, but when they realize you aren't going to hurt them, they will stand still while you scratch them. After a few days of doing this, you can put a halter on the one you are going to train.

## Getting Used to the Halter

If the calf is nervous, put him in a chute to brush and scratch him, then halter him. Use an old rope halter, not your best one. Let him wear the halter and drag the rope for a week in a small pen to get him used to the halter and the pull of the rope. As he steps on the rope, it pulls on his head, and he has to stop.

## Getting Used to Being Tied

After he has worn the halter for a few days and has learned to stop when he steps on the rope, you can tie him up. Wear gloves when handling his rope, so you won't get rope burns if he tries to pull away. Have an older or larger person help you if your calf is still flighty. Tie him in a small pen where you can catch him easily by quietly walking close enough to the trailing rope. Tie him to a sturdy post, not to a pole or rail that would allow him to pull his rope back and forth along the fence. Never tie him to a pole that might pull off the fence. Tie him with his head at eye level, 12 to 18 inches from the post. If you tie him too high or low, he might hurt himself if he pulls back.

## Adjusting the Halter

When putting on the halter, adjust it to fit properly, applying pressure over the nose, not behind the ears. The nose piece should be well up on the nose to prevent slipping. The halter should not be too tight or it will make sores behind the calf's ears.

## Check the Halter

Check your calf's halter each time you catch him to tie him. Make sure it is not rubbing too deeply into the skin over the bridge of his nose, or under his jaw. If it starts to make a sore, readjust the halter or take it off. Start putting the halter on him daily rather than letting him wear it continually.

Tie your calf for only a short time. His early tying sessions should be only about 15 minutes each. As soon as he quits fighting, turn him loose. Leave the halter on. Release him quietly and calmly and don't let him jerk the rope out of your hands. Keep things as relaxed as possible. He shouldn't think he "got away" from you.

Gradually increase the time your calf is tied to about half an hour per day by the end of the first week. Short tie periods help make sure he doesn't hurt himself. If he jerks on his halter for long periods he might strain some muscles or get a swelling under his jaw. Until he learns to behave, always stay nearby during the time he is tied up, in case you have to rescue him.

Don't give him more than 18 inches of rope. If he has too much slack he may get a foot over the rope or his head twisted around, causing him to throw himself on the ground. Always use a quick release knot, and always have a sharp pocketknife handy. If your calf gets in a serious tangle and you can't get him loose, you can cut the rope.

Once your calf settles down while tied, you can gently scratch his shoulder or brush his back and sides. If he is nervous when you are this close, use a show stick to touch and scratch him. This lets you rub him from a safer distance until he relaxes more. Brushing him while he is tied helps calm him.

When you turn your calf loose, pull him around a little with the halter before you let go of him, getting him used to responding to your pull. Daily short sessions are a slow but sure way to halter break a calf. This method is easier and safer than trying to do it all at once.

## Leading Your Calf

After your calf ties well, teach him to lead. Lead him around the small pen for about 10 minutes at a time before turning him loose for the day. When teaching him to lead, use a "pull and release" technique rather than pulling steadily on him. If he comes a few steps as you pull, reward him by slackening the pressure on his halter and speaking to him. When he learns that the pressure eases when he walks forward, he will lead. Talking to him and praising him is important.

When leading the calf, walk on his left side. Hold the lead rope in your right hand, about a foot from his halter. Keep the calf under control, with his head up. If he gets his head down low, he has more leverage to pull harder if he tries to yank away from you.

## Work in a Larger Area

After you've led your calf a few days in the small pen and he leads well, lead him in a larger area. Have someone help you, to tap him on his rear if he balks. If he tries to go fast and won't slow down, use a stick to tap his nose when he speeds up. But don't beat him with it. Never hit your calf hard or use a show stick as a club. Be patient. Try to be a good teacher.

It's important that the calf doesn't learn he can get away from you. Don't try things before he is ready, or he may escape because he hasn't learned to respect the halter yet. Your calf is bigger than you are, so before you try leading him around you must convince him during the tie-up sessions that the halter is the boss.

## Practice Stopping and Turning

Once your calf learns to lead, practice stopping and turning so he'll behave and go the way you want. Try to end each lesson on a good note, quitting for the day while he is behaving and doing things right. His reward will be ending the lesson. If you quit while he is misbehaving, he'll think he can do that and get away with it.

Teach your calf to stop whenever you ask him to. Scratch his belly with a show stick to relax him while he stands still. Some calves resent this touching at first. Wrap the lead rope around a post to keep him in

## How Not to Train Your Calf

Do not tie the calf behind a vehicle to pull him along. The constant pressure on his halter could injure him, break your halter, or cause him to fight it and injure the people around him. Do not beat him with a stick or whip, or pull on the rope with hard jerks. Do not poke him with a sharp object.

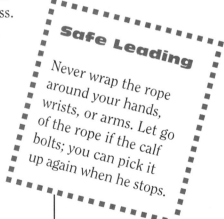

### Safe Leading

Never wrap the rope around your hands, wrists, or arms. Let go of the rope if the calf bolts; you can pick it up again when he stops.

one place when you scratch him the first time. He will eventually come to enjoy this, but at first go easy.

## Setting Up a Calf and Placing His Feet

When your calf is set up in the show ring, the feet should be square under him with a leg under each corner of his body. Don't stand him stretched out too much or with hind feet up underneath him too far; you want him standing in a natural way but squarely. The feet should not be too close together or too far apart.

With practice, the calf will understand what you want. He will usually stand "square" when you halt him, without your having to set him up. He should learn to stand still for 10 minutes at a time, since he may have to stand this long in the show ring.

## Using the Show Stick

Once your calf gets accustomed to being scratched with the show stick, use it to place his feet, to keep his topline straight while standing, to calm him, and to help control him.

When setting him up, switch the lead rope to your left hand and the show stick to your right. If your calf needs a hind foot moved back, pull backward on his halter and at the same time apply pressure with your show stick to the soft tissue between his toes on that foot. (Don't jab! This spot is tender.) If he needs a hind foot moved forward, pull frontward on the halter and use the prong of the show stick under his dewclaw to pull the foot forward. When the hind feet are too close together, apply pressure with the stick to the inside of the far leg, just above the hoof.

Front feet can be repositioned with the show stick or your foot. (When working with a calf, you should always wear boots, not soft-soled shoes!) It's always safest to use the stick, however. When pulling or pushing on the halter, also apply pressure to the dewclaw to move the foot forward, or to the split of the hoof to move it backward or sideways. With practice you can set up your calf quickly and safely.

You can use the show stick to keep the calf's topline straight. When the calf is standing correctly, gently rub under his belly. He will come to associate standing still with the rubbing. It helps calm and relax him and also keeps his topline up and straight. Use slow, long motions with the stick, rubbing his belly in front of his navel area.

## How to Use a Show Stick

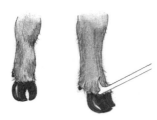

To move the foot back, apply pressure to the soft tissue where the hoof splits; at the same, time push backward on the halter.

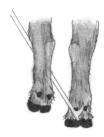

To move a leg forward, pull forward on the calf's halter and put the prong of the show stick behind the dewclaw to pull the foot forward.

Stroking the calf under the belly (in front of the navel area) helps keep him calm.

The show stick can be used while leading to keep the calf from moving too fast. Hold it in your left hand so you can use the butt end for tapping his nose, if necessary. Train the calf to stop by raising his head. If he stops with his head up, he will tend to place his feet squarely.

## Practice for the Show

Your calf should learn to stand quietly while other people walk around him, as the judge will do in the show ring. Have a friend or family member act as judge. Having different people occasionally feel the calf over his back and ribs will get him used to being handled by strangers.

Work with your calf every day. You'll make the best progress if you practice often. Regular short sessions are much better than a few long ones now and then. To get your calf used to sounds and noises, play a radio for him near his stall or pen, starting out softly and then gradually louder as he gets accustomed to it. As he gets gentler, have more of your friends come see him and help you practice, putting on a mock show.

## Care of Your Calf's Hair Coat

A good hair coat helps a calf look his best. No matter what his hair looks like to begin with, you can make it appear better with daily brushing. Begin at least two months before the show.

To groom his coat, tie the calf short (less than 10 inches of rope) so he can't move around a lot. Brush off any dirt, then rinse the calf with your garden hose or wet him with water from a hand sprayer. Then you can start working his hair. It should be brushed and combed forward and slightly up, to give the calf a smoother appearance. Be sure to comb all the hair. Don't skip his legs. The hair over his tail head should be combed straight up. Continue brushing and then combing, finishing with the scotch comb over his whole body.

## Show Equipment

You will need some special equipment to show your calf — a show halter, grooming supplies and clippers, and a grooming (blocking) chute with movable sides. You may be able to borrow some of these things or share equipment with another exhibitor.

Here are some things you need for cleaning and grooming your calf.

Use a leather show halter a few times at home so the calf gets used to having the chain under his chin. For proper fit, the nose piece should be about halfway between the nose and eyes. A well-oiled brown or black leather halter is best for showing; a bright color may distract from the overall impression of your calf.

A good way to train hair is to spend 20 minutes a day brushing and combing. You can't overdo the brushing. It stimulates hair growth and natural oils that make the hair look healthy and shiny.

To encourage hair growth in warm summer weather, keep your calf cool, and increase the number of times per day you wet him down and brush him. During the heat of summer, work with your calf's hair early in the morning and late in the evening, during the coolest parts of the day. In the daytime, keep your calf in a shady place or have a fan by his barn stall. If you are in a 4-H club, members who have experience with brushing and fitting calves can give you advice and show you techniques for working hair.

## Washing Your Calf

Your calf must be clean for showing. Wash him a few times at home, for practice, so he'll be used to it. Washing your calf can be a lot of fun, but it's not a time for games.

Use a halter that won't be ruined by water. Tie him to a secure place, with only a few inches of rope so he cannot move around. Fill a bucket with your hose and add enough livestock soap to make suds. A mild dish-washing soap like liquid Joy or Ivory will work.

Before wetting down the calf, use the scotch comb and a wash brush to remove as much mud, dirt, and manure as possible. Then turn on your hose with low pressure and use your finger on the end to form a spray. Starting at his feet, gently wet his legs completely. He won't like it at first so go slowly, with a gentle spray of water, until he relaxes.

After your calf becomes familiar with having his legs sprayed, slowly wet his belly and work up the body toward his back. Last, wet his head, very gently, holding each ear cupped tightly closed with your other hand.

After the calf is wet, use the scotch comb to remove any mud. Then use a sponge or rag to apply soapy water from the wash bucket to the wet calf, and scrub him with a rubber brush. Wash his legs and belly as well as his sides and back. When you wash his head, again be very careful not to get any soap or water in his ears.

### Rinsing

To rinse the calf, work from the top down. First rinse his head, cupping each ear closed with your hand. Then start at the topline and rinse all the soap out of his hair. Don't leave any or it will dry out the hair and skin and cause dandruff.

Empty the wash bucket and rinse all the soap out. Fill it with clean water. Pour the water over the calf's topline, starting at his shoulders and moving toward

## Never Get Water in His Ears!

Water in a calf's ear will bother him a lot, and the ear will hang down. If water gets deep into an ear, it can cause an infection.

## Be Careful with Soap

Don't put soap directly on the calf. It could irritate his skin and cause scaly skin or dandruff. Just use the soapy water. Use soap by itself only on a persistent manure stain, and then make sure you get it all rinsed out afterward.

his rump. Don't pour any over his head. Then brush and dry him, continuing to brush until he is nearly dry.

Daily rinsing with just a fine spray from a hose or a hand sprayer can help your calf grow a better hair coat. He only needs to be thoroughly washed if he becomes dirty, but wetting his hair can keep him cool and help you with training his hair coat.

## Clipping Your Calf

Clipping is also called blocking. Clipping your calf is fun, and it makes him look nice. The key to a good clip job is knowing your calf's assets and faults. You are trying to enhance his best points and minimize his weaknesses. Imagine yourself as an artist, trying to sculpt the ideal calf.

Your calf should be clean before you clip him. It's not a good idea to wash him just before clipping him, because the hair cannot be smoothly trimmed. Wash him a day before clipping him, and always work his hair just before clipping. Put the calf in a blocking chute or tie him to a sturdy post.

The goal of clipping is to make your calf look smooth and well balanced.

You'll need regular beef clippers such as Sunbeams with standard flat blades for shaving close areas, and plucking blades for where you need to leave more hair. Small-animal clippers with adjustable blade settings also come in handy and are easier to use if you are a beginner with small hands. Your 4-H leader or an older member can give advice on how to use clippers. You may be able to borrow them from someone in your club.

If you are a beginner, use two hands: one to steady the clippers and one to hold them. Trim the hair in an upward and forward motion. Work on one area at a time. Evaluate each section carefully before you start it, and trim the hair as you think it should look. If the calf needs to look more muscled, leave the hair long enough to appear full and rounded. Make a picture in your mind of the effect you want to create. Your goal is to make the calf look smooth and well balanced. You will always make a few mistakes, but don't panic; mistakes are a way to learn. Clipping is a skill developed with practice.

Clipping can be done several times before a show, so you can have more chances to do a better job. Clip your calf just after he is halter trained, getting rid of any extra hair that is not needed during the hair training process. Clip him again about 10 days before the show. Then do the final touch-up job at the show.

## Getting Ready for a Show

Preparations for showing your calf begin well ahead of the show date. Plan ahead to make sure you and your calf are ready.

### Grooming

Wash or rinse your calf once a week beginning a month before the show. He also needs to get used to standing quietly in a blocking chute. This is a portable frame that keeps him in place while you groom or clip him. It is especially useful while working on his leg

## Feet Trimming

Some calves need their feet trimmed before a show. If yours does, have an experienced person do it at least three weeks before the show.

hair. He should get accustomed to calmly entering and leaving this chute, so it's good to practice a few times.

Get him used to a hair drier. This is a handy way to dry him after he has been rinsed or washed, while you comb and brush him. If you blow his hair in the same direction as you brush it, it makes the hair more manageable and easier to work with. Try to use a hair drier on your calf a few times before the show.

## Using the Show Halter

Use your show halter. If it is new and stiff, spray it with leather oil first.

To get your calf used to the new halter, put it over his rope halter and lead him with both halters on. This will still give you control with the old halter, which he is accustomed to, while he is adjusting to the feel of the new one. You can make sure it is adjusted to fit

### Changing Halters

When putting on your show halter, don't take off the rope halter first. Don't take a chance on the calf being completely loose. Put a safety rope around his neck before changing halters.

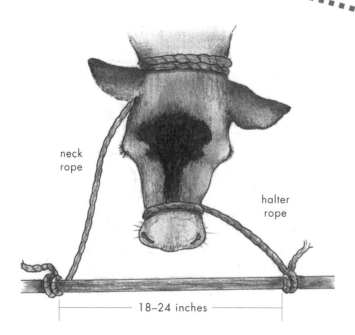

neck rope

halter rope

18–24 inches

When tying with a show halter always use a neck rope, so the calf can't break the show halter if he pulls back. The neck rope should be tied a little shorter than the halter rope so it will tighten before the halter does.

him before you put it on by itself. Then lead him with just the show halter a few times.

Never tie your calf with the show halter; he could ruin it if he pulls back. If you have to tie your calf with his show halter, such as in his stall at the fair, use it with a neck rope (see illustration), or with a nylon halter placed over it. If you use a neck rope or extra halter, make sure it is tied a little shorter than the lead on the show halter. If the calf pulls back, the pressure will not be on the show halter. If you use a neck rope, tie a nonslip knot (such as a bowline) around his neck so the calf cannot be choked.

## Before You Leave for the Show

Make a list of everything you'll need at the show. Gather your equipment and supplies together. Feed your calf only a part ration on the day before you leave for the show. He will travel better than if he is stuffed full of feed.

## At the Show

Arrive at the show with time to spare so you won't be rushed. Unload your calf and tie him while you prepare your stall and put in bedding. After the stall is ready and you've fed and watered your calf, give him a chance to rest and get settled. Be sure to use a neck rope to double tie your calf if he is wearing his show halter.

## Keep a Schedule

If you are at the show for several days, keep a schedule. Exercise your calf in the early mornings before you feed and water him, rinse and work his hair after he has a chance to eat, re-bed his stall in the afternoon, feed and water again in late afternoon, then clean and re-bed his stall in the evening. There is a lot of work to showing a calf.

## Help Your Calf Relax

Being at the show or fair may make your calf very nervous. There's lots of activity around him that he's not used to. Try to keep him relaxed. If he's accustomed to a radio, play it softly in his stall. Talk to him a lot.

## Keep Your Calf Comfortable

Your first priority is to keep him comfortable — providing feed and water and keeping his stall clean and cool. Make him a clean bed with plenty of straw. Wet down the straw with a hose and then pack it into a good bed with your pitchfork. This will help keep him cool. If the weather is hot, set up fans to blow across his back. Wet him down periodically throughout the day.

## Feed and Water

Keep your calf on the same feeding schedule you had at home, if possible. For the first feeding, reduce the amount to about two-thirds of his normal ration, then gradually bring him back to full feed. This will lessen the chance of bloat, a common problem at shows. Observe him closely for signs of sickness. Because of the sudden change in environment, your calf may be more susceptible to digestive problems. Pay attention to how much he eats and drinks.

Give him fresh water periodically. Don't leave a bucket in front of him; he may spill it and mess up the stall. Offer water several times a day, especially in hot weather. Don't use watering tanks where other animals drink. It's safer to have your calf drink out of your own buckets filled at a hydrant. This reduces the chance that he will pick up a disease. Use a bucket that the calf can get his head into clear to the bottom, without spilling.

## Water from Home

Bring some water from home in case your calf won't drink the unfamiliar water at the fair.

## Exercise

Exercise your calf during the cool of early morning or evening. Use a strong halter and rope when leading him. Even though he is tame at home, he might get frightened at the fair. Have an adult with you.

## Proper Restraint

Proper restraint of your calf is important at all times, including proper use of a neck rope and halter. The halter should be tied to the left side of the calf's stall and the neck rope to the right side. This allows him to eat, stand, and lie down without getting tangled up. The neck rope gives added insurance that your calf cannot get loose even if he rubs off his halter.

## Present Your Calf Well

Be polite to visitors who look at your calf, and answer their questions. Enjoy the time you spend with your calf and meeting the people who come by to see him. Keep your stall, your calf, and yourself neat and presentable. Your calf should look his best. Blow the dust or sand out of his hair several times a day and work his hair often. Do your touch-up clipping the day before your class, clipping any hair that has grown too much or that you missed at home.

## Show Day

On the day you show your calf, allow time to exercise, rinse, and dry him. He must be dry before you begin his final fitting. Do your best job of getting him ready, using all the skills you've learned.

## Getting to the Show Ring

Put on your entry number, grab your show stick, and take your calf to the ring, with your scotch comb in your back pocket for touching him up if his hair gets mussed. On your way to the ring, don't let your calf step in manure, or brush against anything that will mess up the hair on his sides. Use any time spent waiting for your class to touch him up with the scotch comb. If you have a chance while waiting, watch the class ahead of yours and see how it is conducted.

## Showing Your Calf

Be courteous in the ring and considerate of other exhibitors. Respond quickly to requests from the judge, but be calm and move quietly with your calf. Even

## Proper Dress

Your appearance will make an impression on the judge, also. Wear clean clothes appropriate for the show ring, such as neat jeans and a shirt with a collar. Faded or frayed jeans won't look good; neither will dirty clothes or tennis shoes. Wear shoes or boots that provide adequate protection in case your calf steps on your foot.

though you are nervous inside, try to show the audience you are proud of your calf. Treat him with care and respect. Try to relax and enjoy this chance to show the judge what you and your calf have learned.

Unless given other instructions, lead your calf clockwise around the circle. This is important to remember in case you happen to be the first one to go into the ring. You may be asked to line up side by side with the others in a straight line facing away from the judge.

## Setting Up Your Calf

Lead your calf into line and set him up quickly without making him nervous. Then turn and face the rear of the calf, holding the halter in your left hand, and the show stick in your right. Leave 3 to 4 feet of space between your calf and the next. This gives both you and the other exhibitor room to work with your calves. Leave at least 5 feet between your calf's head and the edge of the ring, to give the judge room to move in front of the row of animals.

Try to stay in position to watch both your calf and the judge. Keep the calf's legs placed correctly, his head up, and his back level, especially when the judge is nearby. When lining up head to tail, allow 4 to 6 feet between calves. Use patience if your calf is fidgety or difficult to set up. Sometimes it's best to just pull out of line, make a clockwise turn, and start over.

## Parading Your Calf in the Ring

When leading your calf around the ring, walk on his left side with the lead strap in your right hand. Any extra strap can be folded and held in your right hand, or the extra length can be held in your left hand. The latter method is best. You have more control if your calf spooks and tries to take off if you have two hands on the strap. Never wrap the strap around your fingers or hand. Many exhibitors shorten the leather strap so it needs no folding and just hangs free. But this doesn't give you any extra lead if you need it to control and hang onto a startled calf.

## Show Ring Manners

Use good sportsmanship in the show ring. Never lead your calf past the front of other animals or have him out of line blocking the judge's view of another calf. Avoid bumping or crowding other calves.

## Be a Good Sport

You may not win your class, but you can always do a good job of showing. Even if you don't have the best calf, you can have a well-trained, well-shown animal. This in itself makes a favorable impression. Remember to smile! Good sportsmanship is an important part of showing. Sincerely congratulate the winner of your class, or if your calf won, thank everyone who congratulates you.

Carry the show stick in a vertical position in your left hand. Give the calf 6 to 8 inches of lead, holding the lead strap where the leather and chain part meet. If you hold it closer than this, the calf may fight the restraint. If he has more than 2 feet of slack, he will be hard to control. He should lead with his head about even with your right hand and shoulder. Hold his head at a natural level, slightly above the height of his withers. Your hand should be held slightly above the height of his poll.

When leading, leave one to two animal lengths of space between you and the calf in front of you. If the calf ahead slows down or balks, tap him gently to encourage him to move forward. When asked to stop, make sure your calf is in line with the others, then begin setting him up.

### Pay Attention to the Judge

As the judge walks around the class, pay attention. As he approaches, calmly scratch your calf's belly. When the judge moves to the front of your calf, step back a little to give the judge a better front view. Move to the left front of your calf so the judge can handle the right side of your animal. If the judge handles your calf, use your show stick to scratch the calf lightly under the belly to keep him calm.

When the judge moves around in front of your calf, step back to the left side so the judge has a good view. Never get between the judge and the calf; don't block his view. After the judge finishes handling your calf and has moved on to the next calf, use your scotch comb to touch up the hair again.

Be alert and ready to have your calf at attention whenever the judge moves back to your end of the line. Let your calf relax a little in the meantime; if he has to stand in one position for a long time he may get tired. When the calves are lined up side to side, the judge may want to switch the order. Watch the judge and ringman for directions and respond promptly.

## After the Show

After your class, replace your show halter with the rope halter and prepare your calf to be washed. If you used any special sprays or preparations on his hair, this must all be carefully washed out. If you are selling your calf at a fat stock sale at the end of the fair, you will groom him neatly again for that. Selling the calf you have worked with all year is not easy, and you will miss him. But if he does well, you will also be very proud of him.

Once you return home, take time to write thank-you notes to the donors of any awards or trophies you may have won, to the buyer of your calf if you sold him, and to any other people who helped you, such as 4-H leaders. Take time to reflect on your calf project, evaluating the things you did well and the things you need to improve upon. Set some goals for next year. It won't be long before another calf becomes the center of your attention.

**CHAPTER 10**

# Raising a Beef Heifer

**I**f you want to raise a calf you won't have to sell when it grows up, maybe you'd like to have a breeding heifer.

Selecting a heifer for breeding is a little different than selecting a steer to fatten for market. The beef calf is judged on its beef characteristics — lots of muscle and good frame. A breeding heifer should also be able to grow fast. But even more important than her ability to put on weight is her ability to become a mother cow. The biggest, fattest heifer doesn't always make the best cow. Her body is programmed more for getting fat than for getting pregnant or giving lots of milk for a calf.

If this is the first time you've picked out a calf, have someone help you — such as your 4-H leader, FFA teacher, county agent, or someone who raises cattle.

## What Breed?

What breed you choose will depend on your own personal preference, the breeds available in your area, and whether you want a registered purebred, a straightbred, or a crossbred.

A registered heifer may cost more than a commercial heifer. This does not mean she is any better. It just means she has registration papers, which you should receive when you buy her. You could sell her calves as purebreds if you breed her to a registered purebred

*Straightbred. An animal with parents of the same breed, but not necessarily purebred.*

**90**

bull of her breed. To sell her calves as purebreds, you have to register them when they are born. To raise registered cattle, you must join the breed association and pay a fee, and pay registration fees for every calf you register.

If you just want to raise good cattle, producing top-quality market calves, they don't have to be purebred. In fact, the best beef cows are often crossbred, combining the good qualities of two or more different breeds. The choice is yours. Just make sure the heifer you pick has good conformation and feminine characteristics.

## Selecting a Good Beef Heifer

Conformation and disposition are two of the things you should evaluate when selecting your heifer. You'll also want to know her mother's performance in producing calves and her sire's performance in producing offspring. The heifer will be a lot like her parents.

### Performance Records for Purebreds

If you are buying a purebred heifer, use the breed's performance records to help you in your selection. Most successful breeders keep detailed records and use them to identify genetic differences in their cattle. You can use this information to compare things such as birth weight, weaning and yearling weights, milk production, and fertility.

Most breed associations and breeders use EPDs in evaluating their cattle. "EPD" stands for "expected progeny difference." Progeny means offspring; a cow's calves are her progeny. This is a big term that means the cattle are compared against one another to come up with a score to show how they rank in the herd or in the breed as a whole on any specific trait, such as weaning weight. When selecting a heifer as a future cow, you are also selecting the possible genetic traits of all her offspring. The EPD is a tool for helping predict the possible characteristics of her calves.

**Registration Papers**

Registration papers give the pedigree of the registered animal and its registration number, certifying that this animal is recorded in the herd book of its breed.

*Fertility. Ability to reproduce.*

Performance records rank each animal. You can use them to see how a heifer you are considering compares to other animals in the herd or breed. If selecting a purebred heifer, have your 4-H leader or a breeder explain the records, so you can understand how to use these in comparing the animals you look at.

## Choosing a Heifer Without Performance Records

If the calf you choose is not a purebred, you won't have this type of performance record to help you in your selection. But the stockman who raised her can answer questions about the heifer's mother and sire. Was the cow a good mother? Did she calve easily without any help? How many calves has she had? Is she fertile, always breeding early in the season, never skipping a year or calving late? Does she have a good udder? These are important questions to ask.

If buying a crossbred heifer, remember that a good crossbred will usually outperform most purebreds in all traits, because of her added hybrid vigor. But she must have good parents to do this. Just because she is crossbred doesn't mean she will automatically be a good cow. She has to have good genetics. Find out as much as you can about her sire and dam.

## Conformation and Frame Size

Conformation is also important. Cattlemen judge a cow as much by the way she is built as by any other factors. This is just as important as performance records. Your heifer should have good feet and legs. You don't want a cow that will eventually become crippled because of poorly formed feet and legs.

She must have a feminine head and neck, and a long body. You want her calves to be long, because that means they'll be better beef animals. A long body gives a cow more room for carrying a calf. She should have a deep body, not narrow or shallow. She should have muscling, but not the bulging muscles of a steer or bull. If a heifer looks like a steer, she is not a good

*Sire. A calf's father.*

*Dam. A calf's mother.*

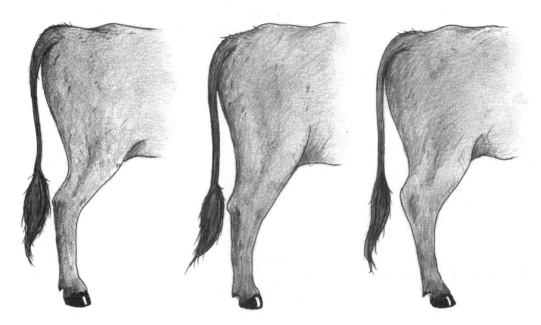

Ideal leg conformation
(from side)

Hind leg too straight
("post legged")

Sickle hocked
(hocks too bent)

Ideal, straight leg
conformation
(from rear)

"Cow hocked"
(hocks too close together)
and splay footed

Feet too close
together

choice for a breeding heifer. She should move freely as she walks, with nice athletic ability. She should have style — all her parts go together to create a good-looking animal.

She should have a good frame, not too small or too large. A really small cow won't raise big calves. A really big heifer may grow into an enormous cow, and it may take too long for her to reach breeding age. Huge beef cows are not very efficient. They take too much feed and they often are not as fertile as they should be. A good medium-sized cow is actually more productive; she'll generally wean a calf that is larger, in relationship to her own body size.

## Udder

You want your heifer to give lots of milk when she grows up, so she can raise a big calf. To do this she must have a good udder. A poor udder might get saggy and become injured, or have big teats or long teats that a young calf would have trouble nursing.

Udder shape and size are inherited. When selecting a heifer, it helps if you know what her sire and dam were like in this area. When selecting a heifer calf from someone's herd, ask to see her mother and also ask about her sire's dam (the mother of the heifer's father). Sometimes you can look at heifer calves before they are weaned, which allows you to examine their mothers.

## Disposition

A cow's disposition is created partly by heredity. She inherits from her parents a tendency toward being nervous or placid, flighty or calm, smart or stupid, kind or mean. Just like humans, some cattle are smarter than others, and some are more emotional. But disposition is also influenced by how the heifer is handled or trained. A timid, nervous heifer that is smart will often gentle down with patient handling. On the other hand, some wild and nervous cattle can be very frustrating, and also dangerous, because they never learn to trust you.

## Good Attitude

A heifer's attitude is very important, especially if you are going to keep her as a cow. You want a smart, gentle cow that will be nice to work with, not a wild, mean cow that could be dangerous.

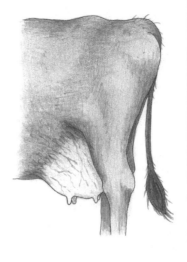

A cow's udder should have short, small teats and well-balanced quarters.

You can't tell by looking at your heifer's udder exactly what it will be like when she matures, but there are clues. Does she have small teats? If her teats are fat or long at this age, they will be even worse when she calves. Poor udders are a serious fault. Many beef cows have to be removed from herds because of this.

## Bad udders

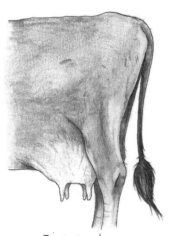

Teats too long

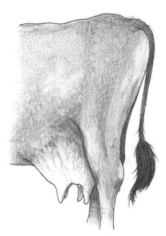

Teats too fat

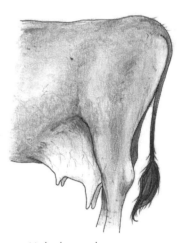

Unbalanced quarters or saggy udder

## Getting Started with a Purebred Heifer

Most breed associations have regional managers who work with cattle breeders to promote their breed, help people get started in the purebred business, and help them improve their herds. They often assist junior members who are starting their own breeding projects. If you contact your breed association, someone will direct you to local breeders who could sell you a heifer. They can also tell you where you can show your heifer if you want to show her in purebred shows and how you can become involved in state and national junior breed activities. Once you buy your registered heifer, give the seller all the information needed to transfer the registration to you.

## Care of Your Beef Heifer

Whether your heifer is a purebred or a crossbred, you will care for her much like you would any other beef calf (as discussed in earlier chapters), except that you will feed her differently. Proper nutrition is important. You must feed the heifer adequately for maximum growth, so she will reach puberty in good time for breeding. But don't overfeed her to the point of fatness, since fat will be detrimental to her.

### Pasture

If you have good pasture during the summer months, this is the ideal feed for a growing heifer. It will save you money because you won't have to buy grain and hay to feed her during this season. Green grass provides all the food ingredients she needs. All you'll have to supply is salt and water.

### Feeding Grain

You shouldn't have to feed your heifer much grain, if any. If she needs grain in order to grow fast enough to meet her breeding weight and mature weight on schedule, she probably won't be a profitable cow. You want a heifer that will grow well and produce good calves without a lot of pampering.

If you are going to show your heifer, you may not be able to win a beef heifer class without feeding grain and getting her fatter than she should be. This is a problem in the cattle industry today. The people who breed purebreds and show cattle and the people who make a living growing beef cattle economically disagree on this issue.

## When You Must Feed Grain

Sometimes you don't have good pasture. Perhaps your area is suffering from drought. Maybe you don't have space to pasture your heifer all summer. There are times you might have to feed hay or grain. If you have good alfalfa hay to give your heifer on poor pasture, or to add to grass hay, you probably won't need grain. But if you don't have alfalfa hay, you can feed grass hay and some grain, and a protein supplement.

If you don't have good pasture, your county Extension agent, 4-H leader, FFA teacher, or some other experienced person can help you figure out a proper growing ration for the heifer.

With proper care and management after she is weaned, your heifer can continue to grow without getting fat. If she weighs between 500 and 600 pounds at weaning and has the genetic potential to weigh 750 to 800 pounds at breeding age (15 months) without being fat, she must gain 150 to 300 pounds in the 160 days between weaning and breeding age. She should be able to do this on good hay. But if you have poor hay, or if your heifer needs more feed to grow this quickly, add a little grain to her ration. If that's what you do, start her on grain gradually, with only a small amount at first.

## How Much Should I Feed Her?

Keep close track of how your heifer looks. Does she appear thin or fat? When in doubt, use a weight tape to see if she is on schedule for reaching her target weight on time. Different breeds have different target weights. The chart on page 67 gives estimates of target weights for some of the breeds and their crosses to give you an idea about the effect that weight has on puberty or the start of heat cycles.

If your heifer isn't gaining weight fast enough, you can feed her more. If she is getting too much feed,

## Keep Your Goal in Mind

When feeding your own heifer, keep your ultimate goal in mind. If you are keeping her for breeding, don't get sidetracked by the short-term goal of showing her. If you want to raise calves, it's best to have her well grown but not fat, even if it means she might not place at the top of her class at the show.

## Weight at Breeding

At breeding age (14 to 16 months), heifers should be about 65 percent of their mature weight. British breeds need to weigh at least 650 to 700 pounds at breeding time, and heifers of larger-framed breeds (such as Simmental and other large European breeds) need to weigh 800 to 850 pounds by breeding time, to be 65 percent of their mature weight.

she'll convert the extra feed into fat. That's a signal that you should cut back on the grain or eliminate it from her ration.

A growing heifer that is too thin will not breed on schedule. She may be late starting her heat cycles in the spring. An overfed heifer will convert the extra calories into fat deposits around her internal organs and in her udder instead of into growth. She'll reach her desired weight faster than normal and become fat. Evaluate her growth and fatness and adjust her feed accordingly.

## When to Breed Your Heifer

Your heifer should be bred in the spring when she is about 15 months old so that she will calve the next year as a two-year-old. A heifer can be bred after she becomes sexually mature and is having regular heat

## Weight of Different Breed Crosses at Puberty

(Heifers are assumed to be at least 13 months old.
Crosses are from Angus or Hereford cows.)

| Breed | 600 pounds percent cycling | 700 pounds percent cycling | 800 pounds percent cycling |
|---|---|---|---|
| Angus | 70 | 95 | 100 |
| Angus/Hereford X | 45 | 90 | 100 |
| Charolais X | 10 | 65 | 95 |
| Chianina X | 10 | 50 | 90 |
| Gelbvieh X | 30 | 85 | 95 |
| Hereford | 35 | 75 | 95 |
| Limousin X | 30 | 85 | 90 |
| Maine Anjou X | 15 | 60 | 95 |
| Shorthorn | 75 | 95 | 100 |
| Simmental X | 25 | 80 | 95 |
| Tarentaise X | 40 | 90 | 100 |

cycles. This is called puberty. Most heifers reach puberty by the time they are 12 months old, but some will be cycling earlier and some will start later if they are slow-maturing individuals.

You don't want your heifer to calve younger than 24 months (two years) old. She isn't mature enough yet to have a calf easily or do a good job raising it and breeding back quickly while she is still growing. And you don't want her to breed late and calve older than 24 months, or she may calve late every year for the rest of her life.

Gestation lasts nine months after a heifer becomes pregnant. You want to feed the heifer properly to have her breed at about 15 months of age. After a 9-month pregnancy she will calve at about 24 months of age.

**Cycling.** *Having heat cycles, which means a heifer is sexually mature and able to breed.*

**Gestation.** *The time it takes for a baby to develop inside its mother's body.*

# CHAPTER 11

# Breeding and Calving

**W**hen it comes time to have your heifer bred, you'll need to take her to a bull or have her bred by artificial insemination (A.I.). It may be easiest to take her to a farm, leaving her there until she comes into heat and can be bred.

## Signs of Heat

A cow or heifer can only be bred when she comes into heat (or estrus). She must be bred at the proper time to become pregnant. If she is to be bred artificially, you must be able to determine when she comes into heat, then have an A.I. technician insert a capsule of semen into her uterus at the proper time.

Determining when she is in heat can be difficult if the heifer is by herself with no other cattle around. Some outward signs of heat include increased restlessness, pacing the fence or bawling, or having a mucus discharge from her vulva. But not all heifers show obvious signs.

If a heifer is living with other cattle, it is easier to tell when she comes into heat. The other cattle will mount her or she will mount them. The hair over her tail and hips may be ruffled from this activity. The easiest way to tell is to put her with another cow, heifer, or steer for a short time. But if you don't have

***Artificial insemination.*** *The process of placing semen that has been taken from a bull into a cow's uterus to cause her to become pregnant.*

When a cow is in heat, other cows try to "ride" her, mounting her and pretending to mate.

any other cattle, it may be simplest to take her where she can be with a bull for one to three weeks until she is bred.

## Importance of Birth Weight for First Calves

For your heifer's first calf, the main thing is to choose a good bull that sires calves that are small at birth. Birth weight is partly determined by nutrition during pregnancy, and whether it is a first calf. But it is mainly determined by genetics. To play it safe, don't use a bull that sires calves that are heavy at birth.

A heifer having her first calf is not as big as a mature cow. If the calf is too large, it may die during birth, or injure the heifer. A big calf usually needs help being born. It may have to be delivered by cesarean section — a veterinarian cuts through the cow's abdomen and into the uterus, takes

### Be Careful Around a Heifer in Heat

When cows or heifers are in heat, they goof around, and "ride" one another, pretending to mate. A pet heifer can be dangerous when in heat, because she may treat you like she would another cow. People have been injured by pet cows trying to rear up and mount them. Be alert to your heifer's moods and behavior. Don't let her start playful actions that could hurt you. Carry a stick to reprimand her, if necessary.

## Artificial Insemination

A large number of cows can be bred to one bull with A.I. The semen is collected and divided into many small portions, and put in tubes called "straws." These are stored in liquid nitrogen, which keeps them very cold (320° below zero Fahrenheit). The frozen straws can be shipped anywhere.

out the calf, then sews the cow up again. This is not the way you want your calf to be born!

## Taking Your Heifer to a Bull

If you bought your heifer from a local cattleman, you might ask him if he would consider putting her with a bull at his place, and what he would charge. Most breeders will be glad to help you, and may not charge a breeding fee.

Ask the bull's owner to keep track of the breeding date so you'll have it for your records. Then you can predict your heifer's calving date the next spring.

If your heifer is a registered purebred and you want a purebred calf you can register, she must be bred to a registered bull of the same breed. If she's a crossbred, or you want to raise a crossbred calf, choose a bull of a different breed, or a crossbred bull.

## Breeding Your Heifer by A.I.

If you can tell when your heifer is having heat cycles, you can have her bred by artificial insemination. Talk to your local A.I. technician about ordering semen from a bull of your chosen breed. There are several breeding services that collect semen from champion bulls all across the country. Some ranchers and most dairymen use A.I. instead of buying bulls. The price of the semen can vary. Some bulls, especially the most popular champions in their breed, are more expensive. You don't need the most expensive semen; any good bull that sires low-birth-weight calves will be fine.

Watch your heifer closely to tell when she comes into heat. She will probably be in heat 12 to 18 hours. Try to spend at least 30 minutes twice a day, morning and evening, watching her for signs of heat. When you see she is in heat, call the A.I. technician.

With your heifer restrained in a chute, the semen is inserted into her uterus through the vagina. With good luck, she will "settle," another term for becoming pregnant. If she doesn't conceive she will return to heat 18 to 23 days later and must be bred again.

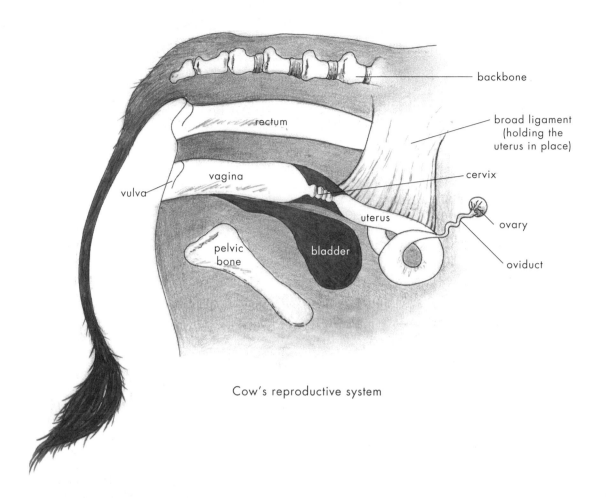

Cow's reproductive system

## Length of Pregnancy

Once your heifer is bred, the fertilized egg will start growing in her uterus. Length of gestation is about 285 days, but she may calve as much as 9 days ahead of her due date or 9 days after. Most heifers calve within three or four days of their due dates.

## Feed During Pregnancy

You won't need to increase your heifer's feed right away just because she is pregnant. Feed as you have been, or keep her on good pasture, so she will continue to grow without becoming thin or fat. If you plan to show her, she should look good, but not fat.

## Showing Your Pregnant Heifer

One nice thing about pregnancy is that your calf won't act as silly as a nonpregnant heifer, and she won't be coming in heat during the fair. This is good, because a heifer in heat may be difficult to handle around other cattle. This can cause the animals around her to misbehave.

## Size of the Fetus at Various Stages of Pregnancy

At two months the fetus is the size of a mouse. By five months it's the size of a large cat. It then grows rapidly, becoming calf sized by nine months.

## Adjust Feed as the Weather Changes

As fall slips into winter, keep close watch on your heifer's pasture. Start feeding alfalfa hay when pasture gets short. Through the winter, feed a mix of alfalfa and grass hay. If she is gaining weight and growing nicely, she won't need grain. Most beef heifers don't need grain during pregnancy, unless they start to get thin. This could happen if the weather is severely cold for a long spell or if the hay is not of good quality.

Feed all the good hay your heifer will clean up. Grass hay provides the roughage and nutrients she needs, while the alfalfa gives the extra protein, calcium, and vitamin A necessary for growth and pregnancy.

If the weather gets cold, increase the amount of feed, especially the grass hay. Grass hay is best for this, since it creates more heat during digestion.

## Feed During the Last Trimester

The last three months of pregnancy, called the last "trimester," is when your heifer needs more feed.

Feed your heifer well, but don't get her fat. If she has a large pasture to move around in, she'll have enough exercise. But if she is confined in a small pen, take her for a short walk every day.

## Vaccinations During Pregnancy

Your heifer should be on a vaccination schedule in which she receives booster shots for certain diseases once or twice a year. Some vaccines can be given during pregnancy and others should not. Talk to your vet about vaccinations your heifer needs.

Also ask the vet about a vaccination to help protect her calf against scours (infectious diarrhea). If your pens or pastures have held baby calves and been contaminated with calf diarrhea, vaccinate your heifer against scours. This will create antibodies against many of the diseases that cause scours. She will pass the antibodies on to her calf when he nurses.

## Get Ready For Calving

As calving time approaches, your heifer will get large in the belly, and more clumsy in her movements. She should be in a safe place where she won't slip on ice or get stuck in a ditch.

Make sure you have a good place for her to calve. A shed in her pen or pasture will work if weather is cold, wet, or windy. If she is confined in a pen or barn, make sure she has clean bedding. The calf must be born in a clean place or he may get an infection.

## Signs that She Will Soon Calve

As your heifer approaches calving, her udder gets full. The vulva, where the calf will come out, becomes large and flabby. Those muscles are relaxing so they can stretch wider when the calf comes through. The area between the heifer's tail head and pin bones becomes loose and sunken. These changes may start several weeks before she calves, or just a few days before.

## Stages of the Birth Process

The process of calving has several stages. Knowing what occurs at each stage will help you decide when or if your heifer needs help.

## Early Labor

The signs of early labor (first stage) are restlessness and mild discomfort. The heifer has a few early uterine contractions as the uterus prepares to push the calf out. She may kick at her belly or switch her tail.

Contractions become more frequent and more intense as labor progresses. The contractions of early labor usually help turn the calf toward the birth canal.

## Things to Have on Hand at Calving Time

- Halter and rope in case you need to tie the heifer

- Strong iodine (7 percent solution) in a small wide-mouthed jar, for dipping the calf's navel

- Towels for drying the calf

- Bottle and lamb nipple, in case you need to feed the calf

- Obstetrical (OB) chains or short small-diameter (½ inch) smooth nylon rope with loop at each end, if you need to pull the calf

- Disposable OB gloves (from your vet) and lubricant (OB "soap") in a squeeze bottle

- Flashlight for checking on your heifer at night

Early labor may last two or three hours in a cow, but four to six hours or longer for a heifer. She gets restless and may pace the fence. If at pasture, she may go into the bushes or some secluded corner.

## Second Stage Labor

When the cervix is fully open and the calf or the water bag — which often is pushed ahead of the calf — starts into the birth canal, active (second stage) labor has begun. The birth should take place in 30 minutes to two hours.

The water sac is dark and purplish. When it breaks, dark yellow fluid rushes out. It may break before it comes out. If this happens, all you'll see is fluid pouring from the vulva. The water sac usually comes ahead of the calf. It should not be confused with the amnion. The amnion is a thin, white sac, full of thicker, clearer fluid. This sac surrounds the calf inside the cow, buffering him while he is in the uterus.

Active labor is more intense than early labor. The heifer has strong abdominal contractions. The entrance of the calf into the birth canal stimulates hard straining. Each contraction forces the calf farther along. Soon the feet appear at the vulva. The calf can safely be in this position for a couple of hours. But it is best if he can be born in an hour. Give the heifer time to stretch her tissues, however. Pulling on the calf too soon may injure her.

Your heifer may get up and down a lot, but once she starts straining hard she will probably stay down. Make sure she doesn't lie with her hindquarters up against the fence or stall wall.

As the cow prepare to give birth, the water sac often emerges first.

It may take awhile to pass the calf's head, but as long as she is making progress you won't need to help. After the head emerges, the rest of the calf usually comes easily. Fluid will flow from the calf's mouth and nostrils as his rib cage is being squeezed through the cow's pelvis. This fluid was in his air passages while he floated around in the uterus. It comes out now so he can start to breathe.

## When the Calf Is Born

After the calf is born, the heifer may lie there a few minutes to rest; labor was a hard job and she may be tired. But the calf must begin breathing immediately. If he doesn't, or if the sac over his head does not break and is still full of fluids, you must quickly help him. Pull the sac away from his nose. Clear the fluid away and make sure he starts breathing.

The cow (she is technically no longer a heifer after she has calved) will probably get up and turn around to see her new calf. She should sniff at him and then start to lick him.

## Shedding the Afterbirth

When the cow gets up after calving, there will be a lot of red tissue hanging down from her vulva. This is the placenta, which surrounded the calf and attached him to the uterus. The attachments ("buttons") are dark red dollar-sized objects spaced over it. It may take 30 minutes or even a few hours for the afterbirth

One of the first things a mother cow will do is to lick the newborn calf.

to completely detach from her uterus and come on out. Cows eat their afterbirth so it won't attract predators. You should remove it from her pen right away so she won't choke on it.

If it takes longer than 10 hours for the cow to shed the afterbirth, she may develop a uterine infection. Keep a close watch for pus discharge or illness, which are signs of infection. If your cow won't eat, or develops a fever, she'll need immediate treatment. Call the vet.

## Helping Your Heifer Calve

Sometimes you need to help with the birth. The calf may be positioned wrong in the uterus and cannot enter the birth canal or come through it. Maybe he's just a little too big to come easily. Or maybe your heifer has twins! If she is too long at labor and nothing is happening, she should be checked. Call your vet or an experienced person to help you. And definitely call for help if you see only one foot, or hind feet, coming out.

### Checking Inside the Cow

When there's a problem with the birth, a careful examination inside the cow may be necessary. If your heifer is gentle, you can check her. First tie her up so she can't move around. If she's lying down and won't get up, check her where she is. Take care to be as clean as possible, to avoid introducing infection into her. Use a disposable long-sleeved plastic glove, if you have one, that covers your whole arm. If nothing has appeared at the vulva yet, feel into the birth canal to see if there are two feet. If there is just one foot, or some other abnormality, you'll know why the birth is not progressing.

If no feet have come into the birth canal, feel farther in and examine the cervix. If it is not opening up yet, you are interfering too soon. When the cervix is completely open, it will be 6 to 7 inches wide and you can reach clear into the uterus.

### Never Pull on the Afterbirth

Never pull on the afterbirth while it is still hanging from the cow. If she doesn't shed the afterbirth for many hours, call the vet.

If the cervix is open and your hand can go through, reach into the uterus and check the calf. You'll probably feel the feet. But if the calf is not aimed right, the first part of him you touch may be his head, tail, some other part of his body, or just one foot. If the calf is not aimed right, you will need help immediately to reposition the calf so he can be born.

## Pulling a Calf

Often the only problem is that the calf is a little too big and needs a pull. But don't pull on a calf unless he is in perfect position to come. If the feet have been showing for an hour and you've felt inside the vulva to make sure the nose is right there and the head is coming properly, have your parents or someone else help you pull the calf.

If the calf's nose is showing and the heifer's straining starts to push the head on out, you can wait. But if she isn't making progress after the feet have been there an hour, you should help her. First, feel inside the birth canal to see if there is room for the head to pass through the pelvic opening of the heifer. If you cannot get your fingers between the top of the calf's head and the top of the birth canal, the opening may be too small. If that's the case, call your vet.

But if you think the head can come through, go ahead and pull on the calf's legs. It helps if there are two people to work as a team to pull the calf. Pull alternately on one leg and then the other, to ease the calf through the pelvis one shoulder at a time.

The calf has to come out in an arc. When his feet emerge from the vulva, you should pull straight out.

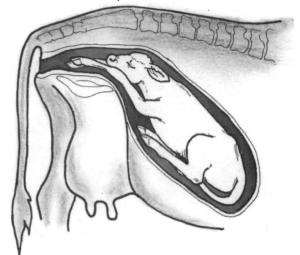

Normal birth presentation

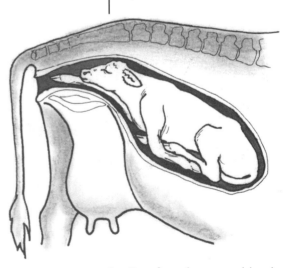

One front leg turned back; the calf must be pushed back and the leg brought into the birth canal.

But after the head comes out, pull slightly downward, more toward the cow's hocks, as his body arches up over the pelvis and then down. If you watch a normal birth, you'll notice the calf curves around toward the cow's hind legs as she is lying there and he slides out.

In a difficult birth, one person can pull on the front legs with obstetrical chains or ropes around the calf's legs above the fetlock joints, so they won't injure the joints or feet. At the same time the other person stretches the cow's vulva (see drawing). This helps the head come through more easily. One person pulls while the other stands beside the cow, if she's up, or sits beside her hips, if she's down, facing to the rear. If you are the one doing this, put your fingers between the calf's head and the cow's vulva, pulling and stretching the vulva each time the cow strains. You and your partner should pull and stretch the vulva only when she strains, and rest while she rests. Don't pull when she is not straining.

You can help deliver a calf by stretching the vulva.

## Hiplock

Sometimes in a hard birth, you get the calf partway out, only to have him stop at the hips. Don't panic. Remember that he has to come up and over the pelvic bones in an arc. As his body comes out, you should start pulling downward, toward the cow's hind legs. To avoid hurting his ribs, get him out far enough so that his rib cage is free before you pull hard downward. If his rib cage is out, he can start to breathe if the umbilical cord pinches off.

If the cow is standing, pull straight downward and underneath the cow, pulling the calf between her hind legs. This raises the calf's hips higher, to where the pelvic opening is the widest. If the cow is lying down, pull the calf between her hind legs, toward her belly.

## Backward and Breech

A calf coming backward, with hind feet protruding from the vulva, has his heels up. Front legs have the toes pointing down. If the bottoms of the feet are up, the calf may be backward. But before you assume that, feel inside the birth canal to see if there are knees (front legs) or hocks (hind legs). The calf may be just rotated a little sideways or upside down.

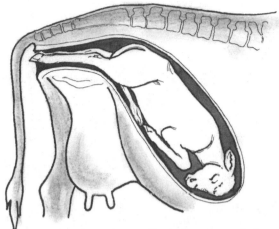

Posterior presentation. Birth is usually too slow to result in a live calf unless assisted.

If the calf is backward, call the vet quickly to help you. He will use a calf puller to get the calf out before it suffocates.

A breech calf is positioned backward, but the legs do not enter the birth canal; he is trying to come rump first. The cow may not start second stage labor at all. Nothing is in the birth canal to stimulate hard straining. She seems to be too long in early labor. If you wait a lengthy period before checking, the placenta will eventually detach and the calf will die. If you check inside her, all you'll be able to feel is the calf's rump or tail. Call the vet.

## Leg Turned Back

Sometimes one front leg will be turned back. One foot will appear but not the other, or sometimes the head and one front foot will show. It's best if you can detect this problem early, before the head is pushed out very far, so the calf can be pushed back into the uterus, where there is room to rearrange him and get the other leg unbent and coming properly. Get help immediately.

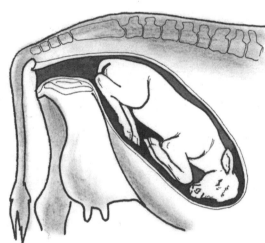

Breech. The calf must be pushed forward far enough so that each hind leg can be tightly flexed at the hock and brought into the birth canal.

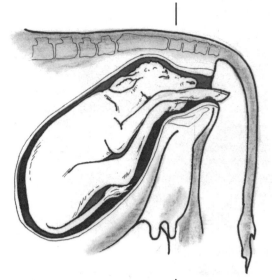

All four feet coming into the birth canal. The calf must be pushed back far enough to push the hind legs back over the pelvic brim and into the uterus.

## Get the Calf Breathing

After any difficult delivery, make sure the calf starts breathing as soon as possible. Stimulate him to breathe by sticking a clean piece of straw or hay up one nostril to make him sneeze and cough. If you get no response, give him artificial respiration. If he is still alive, you can feel his heartbeat near the rib cage, left side, behind his front leg. Blow a full breath of air into one nostril, holding the other nostril and his mouth closed with your hand. Blow until you see his chest rise. Then let the air come back out on its own. Blow in another breath, and keep breathing for the calf until he regains consciousness and starts breathing on his own.

Some calves are still encased in the amnion and its fluids after sliding out. If the sac doesn't break, the calf

To get a newborn calf breathing, tickle her nose with a piece of straw.

will die. Calves won't breathe until the fluid is away from their noses. This is a very good reason to be there when your heifer calves!

## When in Doubt, Get Help

This chapter is not meant to scare you, but to let you know of problems that could happen. Try to recognize problems early and get help before the cow or calf is in serious trouble. It's best to call the vet when you are in doubt about a situation or unsure of your ability to handle it.

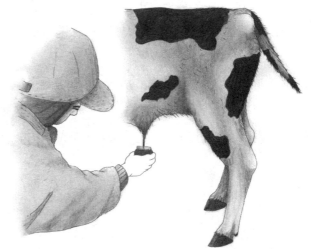

To disinfect a newborn's navel stump, use a small jar containing strong iodine.

## Disinfect the Navel Stump

Disinfect the calf's navel stump as soon as the umbilical cord is broken and you have made sure the calf is breathing. (See pages 117–118 for a description of navel ill.) Have a small wide-mouthed jar (such as a baby food jar) ready ahead of time, with ½ inch of strong (7 percent) iodine in it. Immerse the navel stump in the iodine by holding the jar tightly against the calf's belly and making sure the navel cord is thoroughly soaked in the iodine.

## Make Sure the Calf Nurses

Watch to make sure the new calf is able to get up and nurse.

If the calf doesn't get up within 30 minutes, help him stand. Make sure he nurses within an hour of being born, the sooner the better. If the cow won't stand still while he tries, help him. If she is too fidgety, you may have to feed her some alfalfa hay to encourage her to stand still, or even tie her up.

### Be Careful with Iodine

When dipping the newborn calf's navel in iodine, do not spill any on your hands or the calf. It is a strong chemical and can burn the skin. Never get any in your eyes or in the calf's face or eyes.

## Colostrum

Colostrum, the cow's first milk, is special. It has twice the calories of ordinary milk, with rich, creamy fat that is easily digested and high in energy. It also helps the calf pass his first bowel movements. Of most importance to the calf are the antibodies in the colostrum that give him immunity to disease. This temporary immunity lasts several weeks, until his immune system starts making its own antibodies. A calf that gets no colostrum or that doesn't nurse until he is several hours old may develop scours or pneumonia.

The best time for the calf to absorb antibodies is during the first two hours after birth (preferably the first 30 minutes!). Make sure he is up and nursing within an hour of birth. If he isn't able to do it on his own within that crucial time, help him, or milk a little out of your cow's udder and feed it to him with a bottle.

## If the Heifer Won't Cooperate

Most heifers are good mothers. But sometimes a heifer is confused and doesn't want to mother her calf, or won't let him nurse because her udder is sore. If she kicks at him and he cannot nurse, tie her up or restrain her in a stanchion and tie a hind leg back so she

## Avoid Distractions

Don't have a bunch of people around or invite friends over to see the new calf right away. This could make the heifer nervous, and she might not mother the calf. Wait a few days to show off the new baby. Let Mama get acquainted with her baby without any distractions.

Cow tying can prevent a cow from kicking her newborn calf.

You may have to put hobbles on the hind legs of a mother cow if she continues to kick her calf (probably because the nursing is painful).

can't kick. Leave enough slack in the rope so that she can stand comfortably on that leg, but not enough to kick. Then you can help the calf without her striking out at him or you. Once he has nursed, the calf will be more accepted by his mother.

Sometimes, however, the mother keeps kicking. She may have a lot of swelling in her udder (called "cake"). Nursing is painful for her. You may have to put hobbles on her hind legs for a few days so the calf can nurse without being kicked.

### If a Calf Can't Nurse

If the newborn calf has trouble nursing or is too weak to stand and nurse after a difficult birth, you can milk colostrum from his mother to feed him. You'll need about a quart. Pour it into a small-necked bottle with a lamb nipple and feed it to the calf. This will usually give him strength to get up and nurse.

If he is too weak to suck a bottle for his first nursing, have your vet or an experienced person show you how to use an esophageal feeder.

### From Sassy Calf to Devoted Mother

Now that her baby is safely born, your young cow is probably quite proud of him. Most first-calf heifers are good mothers and take their new job very seriously. It's awesome to see the change when the sassy heifer becomes a devoted mother, licking her calf and mooing at him. She wants to protect him from any danger. She may become dangerously aggressive for a few days. Be prepared, just in case.

CHAPTER

# Managing Your Own Herd

**T**his chapter will look at managing a group of cows, and the various things involved in taking care of them throughout the year. These guidelines will be useful even if you just have one cow.

## Winter

At this time of year, any calves from last spring are already weaned and vaccinated and growing nicely on good feed. If you have the pasture space, your calves will do better on pasture than confined in a corral for winter. A pen will get muddy, and may be knee deep in mud and manure by spring. Calves can get cold standing in mud to eat. Make sure they have a dry, clean place to sleep.

Your cows and yearling heifers are pregnant. Your task is to get them through winter in good shape for calving. Young cows that are still growing need alfalfa. Mature cows may get by on good grass hay. Make sure water supplies are adequate and not freezing up. Break ice often or use a water tank heater.

## Delouse

Lice multiply swiftly in cool or cold weather. Delouse all cattle in late fall and again in late winter. Talk to your vet about a good control program.

## Winter Diseases

Keep close watch for any other problems such as foot rot, or weather-related illnesses such as pneumonia. Your cows and pregnant heifers should already be vaccinated for lepto (and redwater, if it's a problem in your area). If you are also vaccinating the cows to create antibodies in their colostrum to protect their newborn calves against scours, the time to give these shots is several weeks before calving. Heifers expecting their first calves will need two shots, a few weeks apart. Talk to your vet well ahead of time and figure out a vaccination schedule.

## Spring

This is the time of year when your calves will be born.

### Get Ready for Calving Time

If you have several cows or heifers, you will want to tag the calves when they are born. Before calving time, buy your tags and number them. Also purchase any medications and equipment you might need, and have them on hand: iodine, vaccine for newborn calves, obstetrical (OB) gloves, scour medications, and so on. Then if a cow calves a week early, you'll be prepared.

### Prevent Navel Ill

Navel ill is a serious infection that can kill or cripple a calf. Bacteria that enter through the navel may create an abscess in the navel area or may get into the bloodstream and cause a general infection called septicemia, which can be fatal. Or the bacteria may settle in his joints. It can be very hard to save a calf once he gets navel ill, even with diligent treatment.

Have a clean, dry place for each cow to calve, with clean bedding. Disinfect each newborn calf's navel stump with strong iodine. The iodine not only kills germs but also acts as an astringent, shrinking the tissues and helping the navel stump dry up quickly and seal off so bacteria cannot enter the calf. Don't touch the navel cord with your hands unless they are really clean. The only time you need to is if the cord has broken off really long and is dragging on the ground when the calf is standing up. Then you should cut it with very clean, sharp scissors, leaving a 3-inch stump and being careful not to pull on it. If you pull or jerk on the cord, it could injure the calf internally. As soon as you cut it, immediately soak the navel stump in iodine.

The navel stump of a baby heifer usually dries up after just one application of iodine. But a bull calf's stump may take longer. He urinates close to the navel and if he urinates while lying down, as many baby bulls do, he keeps wetting the navel stump. It doesn't dry up very fast. While it is still wet, bacteria can enter it. So reapply the iodine a few hours later, and again if necessary — to get the stump dried up within the first 24 hours after birth.

## Medications for the New Calf

Give the new calf an injection of vitamin A, if recommended by your vet, and any other necessary medications. In some areas, newborn calves need selenium to prevent "white muscle" disease. You may need to vaccinate newborns against enterotoxemia (a highly fatal gut infection) or tetanus. So discuss this with your vet ahead of time.

## Mothers and Calves

Keep each cow-calf pair by themselves a day or two before they go out with other cattle. Then the calf will know who his mother is. He won't get confused and try to nurse the wrong cow and get kicked.

Keep the cows with babies in a different place from the pregnant cows. They need to be fed differently. The mother cows need more feed so they can produce milk for their calves.

## Watch for Scours

The bacteria that cause scours may be lying around on the ground, just waiting for the right conditions, such as wet, muddy weather. The bacteria got there from sick calves in earlier years. Sometimes bacteria are brought into a pen or pasture with purchased calves, or carried from a neighboring farm on the feet of birds, animals, or people.

## Preventing Scours

The key to preventing scours is good management — a healthy cow herd, uncontaminated areas for calving, clean bedding for cows so they don't get their udders dirty, and making sure every calf gets an adequate amount of colostrum soon after he is born. Prevaccinating cows ahead of calving can prevent some types of scours, but not others.

*Cleanliness.* Wash the teats of any cow or heifer you have to assist in calving, clean all your equipment between calves (especially bottles and nipples, and your esophageal feeder tube), and move cows and calves to a clean pasture after they are well bonded.

*Separate sick calves and pregnant cows.* Never put sick calves in the same barn stall or pen where you will have cows calving. If you have sick calves that need shelter, use a different shed. Keep pregnant and calving cows separate from already-calved cows. If a pregnant or calving cow is in a pasture with cows and calves, and she lies in a bedding area where a calf has scoured, she may get these infective germs on her udder — and her newborn calf will get them as soon as he nurses.

*Colostrum.* Make sure every calf has colostrum within two hours of birth. Freeze extra colostrum for emergencies. If a gentle older cow has a lot of colostrum when she calves, you can milk a quart from her while her own calf is nursing. You can freeze it in a plastic container or milk carton. Then you'll have it if you ever need it; frozen colostrum will keep for several years.

Young calves need to be protected from cold, wet weather in order to stay healthy.

## Keeping Colostrum

Extra colostrum will keep for up to one week in the refrigerator, and for several years in the freezer.

When thawing colostrum, don't thaw it in a microwave or get it too hot; excessive heat destroys the antibodies. Put the container in hot water to let it thaw, and never heat the colostrum much higher than calf body temperature. It should feel comfortably warm, but not hot, on your skin before you feed it to a calf.

*Protecting from weather.* Young calves need protection from bad weather. If they get wet and cold in a spring storm, they are more susceptible to scours and pneumonia. You can make a small shed where the calves can get in but the cows can't. Put clean bedding in regularly.

### Doctoring Scours

With many types of scours, a month-old calf may recover quickly, while the same infection might kill a week-old calf unless you give him intensive treatment.

*Antibiotics.* When using antibiotics to treat scours, it's best to give a liquid by mouth rather than shots or pills. Injected antibiotics don't help scours much. You need the antibiotic to go directly into the digestive tract, not the muscle. And pills do not dissolve quickly enough in the stomach. The calf's digestion is messed up and pills are not absorbed well. A liquid antibiotic is much better. Get a good antibiotic from your vet.

*Replace fluids.* The most important treatment for diarrhea is to replace the fluids and body salts that the calf is losing, so he won't become dehydrated. The sick calf needs lots of fluids. You should give him a quart of warm water, by esophageal feeder, every six to eight hours until he starts to recover. Have an experienced person show you how to put the feeding tube down the calf's throat. And every time you give him warm water, add some electrolyte salts to it. You can buy packets of electrolyte mixtures from your vet. You can also use a simple homemade mix.

To make your own electrolyte mix for a scouring calf (one dose), mix ½ teaspoon of regular table salt (sodium chloride) and ¼ teaspoon of Lite salt (sodium chloride and potassium chloride). If your calf is critically ill, add ½ teaspoon of baking soda (sodium bicarbonate). If he is weak, you can add a little bit of powdered sugar (tablespoonful) to the mix to give him energy.

Put it all into some warm water (1 quart for a small calf, or up to 2 quarts for a large one). Add a liquid antibiotic (such as neomycin sulfate solution, sometimes called Biosul — which you can purchase from your vet), using the proper dosage for the size of the calf (follow directions on the label). Give this mixture by esophageal feeder. If you wish, you can also add 2 to 4 ounces of Kaopectate (¼ to ½ cup) to this fluid mix. The Kaopectate helps to soothe the gut and slow down the diarrhea. Instead of the Kaopectate you could use a human adult dose of Pepto Bismol.

## Learn to Use an Esophageal Feeder

An esophageal feeder is a handy way to get fluids into a calf that has scours or pneumonia, or into any sick calf that will not nurse. It is also a way to get colostrum into a newborn calf that for some reason cannot nurse. The esophageal feeder is a container attached to a tube or stainless steel "probe" that goes down the calf's throat and into the stomach. Get one from your vet, and have him show you how to use it.

## Don't Overdo Antibiotics

If you give a calf antibiotics for more than two or three days, it may kill off his good "gut bugs" as well as the bad ones. You may have to give him a pill or paste containing the proper rumen bacteria (obtained from your vet) after he recovers, to restore proper digestive function.

If you catch scours early enough, you can often halt the infection before the calf needs fluids and electrolytes. Neomycin sulfate solution can be put into a syringe, to squirt into the back of the calf's mouth. Use 1 cc per 40 pounds, which means about 2 to 3 cc for a young calf.

## Detecting Illness

In order to treat calves at the first hint of sickness, you need to watch them closely.

There are many clues that can help you spot trouble before a calf is critically ill, besides a messy hind end. Often the calf will be a little dull even before he shows diarrhea. Things to watch for are calves that quit nursing, or a calf lying down and off by himself when the rest of the babies are running and playing. Feeding time is a good time to check on calves and observe them for signs of illness. Also check the mothers' udders. Often the first sign of trouble will be a calf not nursing because he doesn't feel good. If any cow has a full udder or is only partly nursed out, take a closer look at her calf.

## Vaccinations

The cows need to be vaccinated after calving and before rebreeding, for lepto, IBR, BVD, etc. Vaccinate at least three weeks before rebreeding. Some of the live virus vaccines (such as IBR-BVD) may cause abortions in cows if given while they are pregnant. Also, you want the cows and heifers to have time to build immunities against these diseases before they become pregnant, since some of these diseases can cause abortions.

Your young calves need to be vaccinated against blackleg, malignant edema, and other clostridial diseases.

## Castrating

Baby calves should be castrated and dehorned at an early age — the younger the better. All bull calves should be castrated, unless they will be used for breeding.

Castration is harder on a calf the older he gets. As the testicles grow, the blood vessels supplying them also become larger. There is always danger of bleeding when castration is done surgically. This should be performed by an experienced person.

Baby calves are easy to castrate, however, by using elastrator rings (strong rubber rings, about the size of a large Cheerio). These are stretched, using a special tool, and placed over the scrotum, at the top of the scrotal sac, above the testicles. Then the tool is removed, leaving the rubber ring to constrict tightly, cutting off all circulation and feeling below it. The testicle tissue dies, and the scrotal sac shrivels up and drops off a few weeks later, leaving a small raw spot that soon heals.

This is the easiest way to castrate baby calves, and also the safest, for there is no bleeding. To put the rubber ring on a calf, he should be lying down on his side, with someone holding his head and front leg so he can't wiggle around while another person puts on the ring.

## Dehorning

If your calves have horns, they should be removed; cattle with horns are mean to each other and could also hurt you. Dehorning should be done when the calf is young and the horns are small. Older calves or mature cattle with large horns may bleed a lot when their horns are removed, or get infections.

There are several methods of dehorning. A caustic paste or stick can be used for young calves a few days old. Another way to dehorn baby calves is with a battery-operated dehorner.

Electric dehorners are another method, and are often used on calves up to three months of age (when horns are larger than button stage). The hot dehorning iron is held firmly against the head, over the horn, long enough to kill the horn-growing cells at the base of the horn. Have an experienced person dehorn your calves.

## Summer

Your chores will get a little easier in summer if your cows are on pasture. Calving is over and you aren't feeding hay. You'll still need to watch your cattle closely, to be able to treat any health problems that might occur, such as pinkeye or foot rot. Summer is

### Store Medical Supplies Safely

Make sure all live-stock medications and treatments (iodine, insecticides, antibiotics, etc.) and needles and syringes are in a safe place, away from little brothers and sisters.

## Remove the Bull from the Herd

After the cows are bred, take the bull out of the herd and keep him in a separate pen. Bulls can be a nuisance and should not be kept with the cows all year round. Some bulls are mean and most are obnoxious, rubbing on fences and trying to get out and go fight other bulls or find other cows. Bulls can be very hard on fences.

also the time for fly control. You may want to use insecticide ear tags.

## Breeding the Cows

If you are breeding your cows by A.I., watch them for signs of heat. Check for heat twice a day, early morning and late evening. For best results, have the technician inseminate each cow about 12 hours after heat is first noticed. It is standard practice to breed a cow in the evening if she was noticed in heat that morning, or early the next morning if she was seen coming into heat in the evening. Keep good records on the dates each cow has heat cycles, when she was bred, and the expected calving date.

If using a bull, he should be turned in with the cows nine months ahead of the date you want the cows to calve — May 1 if you want them to start calving in early February, or the end of May if you want them to calve in March. If you have heifers, keep them separate from the cows and breed them to a bull that sires small calves at birth. You can use the same bull for cows and heifers if you know he sires small calves that grow fast.

If you only have a few cows and don't want to keep a bull, you can borrow or lease one for the short time you will need him. Forty-five to 60 days is long enough to leave a bull with the cows if they all calved about the same time and none are really late.

## Register Your Calves

If you are raising purebreds, now is the time to get your calves registered if you haven't already; this should be done before they are six months old. Send the information on each calf to the breed association, along with the registration fees. The older the calf, the more it costs to register him, so get it done now.

## Fall

Fall is the time for weaning calves and getting ready for winter.

## Vaccinations

Vaccinate all heifer calves for brucellosis (Bang's) if they haven't already been done. Cows should be revaccinated for lepto and any other diseases your vet recommends. All calves should be revaccinated for blackleg and malignant edema, and any other clostridial diseases in your area, and for viral diseases such as IBR, BVD, PI3, etc. They should be dewormed and deloused.

## Sell Steers and Heifers

Decide where to sell your steers and any heifers you don't plan to keep. Do you want to sell them through an auction or directly to a feedlot? Or to a neighbor who is buying weaned calves to put on pasture? If you aren't sure how to market your calves, talk to your county Extension agent or to local farmers. If you have purebred calves, get advice from your regional breed association representative.

If you sell your calves to a buyer who will be sending them to a feedlot, wean them a few weeks before they are sold, to have them over the stress of weaning.

## Weaning

Weaning is a traumatic experience for a calf, and also for his mother if this is her first calf. Older cows are often not as upset; they've gone through it before.

The emotional stress of weaning is harder on the calf than the sudden lack of milk. Separating calves from their mothers and putting them in a corral by themselves causes much anxiety. They frantically miss their mothers and the security of the herd. Their desperation is contagious. The calves mill around and pace the fence, bawling and running, stirring up dust, which can irritate their lungs and make them vulnerable to pneumonia. Any frantic activity by one sets off a reaction among the others. They all start bawling and pacing again, rarely taking time to eat.

It's easier on calves if they can be weaned in a grassy pasture rather than a dusty corral. If you have

## "Babysitter" Cow

Another way to wean is to leave a "babysitter" cow or heifer with the calves when they are separated from their mothers. She is a calming influence and helps them get through this emotional time.

just a few cattle, a good way to wean them is with a net wire fence dividing a small pasture; put the cows on one side and the calves on the other. The calves will still bawl and pace the fence, but they can see their mothers and be near them, and after a few days they are not so worried.

Make sure your fences are strong when you wean calves. They will try to crawl through, and so will some of the mothers.

### Buying and Storing Hay

Make sure the hay you buy is good quality. Look at the hay before you buy it. You don't want rained-on moldy hay or dry dusty hay. You might want to purchase hay during the summer, right from the field, and have it hauled home. If you don't have room to store a lot of hay, you may be able to make arrangements with the seller to store it for you.

If you store it at home, stack it in a high, dry spot where the bottom bales won't get wet. Cover the stack with tarps to keep the top bales from spoiling due to rain and melting snow.

This haystack has an extra-high row of bales down the center so that the tarp is sloped, allowing moisture to run off.

## Culling Cows

Have your vet pregnancy test your cows. You should probably sell any older cow if she did not settle. It is usually best to cull any cows that are open, or really old with poor teeth that are likely to become thin over winter, crippled, or with a bad udder or any other serious problem.

## Taking Stock of the Year Just Past

Fall is a good time to look back at the year just past, to see how the calves grew, how outstanding the heifers are this year, how the problems and challenges along the way were overcome. It makes you feel good if you were able to save a calf that might otherwise have died at birth, or if you kept a sick one alive with good doctoring. It warms your heart to see him now, big and sassy. It makes you eager to see next year's babies — ready to face any challenges that may come along.

**Cull.** *Remove a low-quality cow from the herd by selling her.*

**Open.** *Not pregnant.*

# PART THREE

# Raising a Dairy Heifer

# 13

# Selecting and Buying a Dairy Calf

**I**f you are going to raise a dairy heifer, you will get her as a baby calf, younger than a weaned beef calf.

A baby calf is fun, and can be easier to tame than a big, weaned beef calf, especially if you are a small person or a beginner. A baby dairy calf is both easier and harder than a beef project. It's easier because the calf is smaller at first, and simple to handle. But raising a baby calf is also a lot of work.

The baby calf is more susceptible to a variety of diseases. An older calf has already developed some immunities. But if you take care of her well, the baby calf will usually stay healthy, and will be a sassy character that looks forward to seeing you. You will be the substitute "mother" for awhile.

## Why Raise a Dairy Calf?

Perhaps you are getting a calf for a 4-H project, or want to raise a heifer to keep as a family milk cow or start your own dairy herd. Or maybe you want to raise a dairy heifer to sell.

You can make money raising dairy heifers. A good young milk cow is worth more than a beef cow; a dairy cow makes more money producing milk than a beef cow can make by producing beef calves.

## Buying a Dairy Calf

Most dairy farms have many very young calves to sell in the spring. Baby calves can be purchased quite inexpensively, compared with a big weaned beef calf. But a good dairy heifer will not be as cheap as a new-born bull calf. Bull calves are cheaper because most dairy people don't want to take time to raise them.

## Why Do Dairies Sell Young Calves?

Dairy cows must have a calf every year to produce their maximum amount of milk. A cow makes much more milk after she freshens. Her volume of milk is greatest a month or two afterward. From then on, her production gradually declines. So to keep dairy cows at maximum production, the cows are bred every year to have new calves and allowed to "dry up" briefly before the new calves arrive.

Dairies have lots of new calves every year. They may keep the heifer calves to grow up to be milk cows, but they don't need the bull calves. Some dairies sell all their calves. Others keep their heifers, and raise them to sell some to other dairies. Bull calves are usually sold as soon as they are born.

Some of the calves at a dairy may be crossbred (half beef), if the dairyman breeds his heifers to a beef bull that sires small calves, for easy calving. Crossbred calves can often be purchased cheaply. They make good bucket calves to raise for beef. A crossbred heifer can become a good family milk cow or nurse cow, but she wouldn't give enough milk to be a good dairy cow.

## Buying at an Auction Is Risky

Some dairies sell calves at an auction. In areas of the country where there are lots of dairies, there are always baby calves at the auction sales.

An auction is the riskiest place to buy a young calf. Even though the calf may have been healthy when she was taken to the auction, she may get sick after you

**Freshen.** *To give birth to a calf and start producing milk.*

## Human Food

Cow's milk has been used for human food for more than 11,000 years. Soon after humans domesticated cattle, they figured out how to milk them. Early pioneers brought milk cows with them when they sailed to America, and took their cows along as they moved West.

## Try to Buy at a Dairy

Check the local dairies to find out which ones would let you come pick out a new calf. If you buy your calf at a dairy, you can ask questions and find out a lot more. This is especially important if you are going to raise a dairy heifer. You'll want to know about her sire and dam. And you can make sure you only buy a calf that has had colostrum and is off to a good start.

bring her home. This is because some calves are taken from their mothers and sold before they have had a chance to nurse enough colostrum. They don't have antibodies against diseases.

A sale yard is also a good place to pick up diseases. Cattle come and go, and spend time in those pens before being sold. Some of the cattle brought to a sale may be sick or coming down with an illness. Even if most of the cattle that go through the pens are healthy, there is a chance for germs to contaminate these pens.

Don't buy a calf at an auction if you have other options.

## Sources of Help with Your Dairy Heifer

A good calf, well raised, will become a good cow. The calf you start with may become the foundation of your future herd. You might want to join a 4-H club or youth program in a dairy breed, or a local breed club, or an FFA program, to learn more as you go along. The 4-H dairy project, for instance, will give you help in feeding and caring for your heifer, and outline the chores that must be done properly to raise your calf to become a good cow.

Leaders in these groups can answer your questions and help with your heifer if you have problems or need advice from experienced people. They can also help you in selecting your first heifer.

Guernsey

## What Breed?

You can be successful with any of the dairy breeds, but you may want to choose one that is popular in your area, especially if you want to sell your heifers.

### Guernsey

Guernseys are fawn and white, with yellow skin. The cows weigh 1,100 to 1,200 pounds. Bulls weigh 1,700 pounds. Guernsey cows have good dispositions and very few calving problems. Their milk is yellow in color and very rich in butterfat. Heifers mature early and breed quickly.

### Ayrshire

These cattle are red and white. (The red can be any shade, sometimes dark brown.) The spots are usually very jagged at the edges. Cows are medium sized, weighing 1,200 pounds. Bulls weigh 1,800 pounds. Cows are noted for their good udders, long lives, and hardiness. They manage well without pampering. They give rich, white milk.

### Holstein

Holstein cattle are black and white or red and white. They are large; cows weigh 1,500 pounds and bulls weigh 2,000 pounds. A Holstein calf weighs about 90 pounds at birth. The cows produce large volumes of milk, low in butterfat. Holsteins are the most numerous dairy breed in America.

### Jersey

Jerseys are fawn colored or cream, mouse gray, brown, or black, with or without white markings. The tail, muzzle, and tongue are usually black. They are small cattle. Cows weigh 900 to 1,000 pounds and bulls weigh 1,500. Jerseys calve easily and mature quickly, and are noted for their fertility. Jerseys produce more milk per pound of body weight than any other breed, and their milk is the richest in butterfat.

Ayrshire

Holstein

Jersey

Brown Swiss

Milking Shorthorn

## Brown Swiss

Brown Swiss are light or dark brown or gray. They are large cattle. Cows weigh 1,400 pounds and bulls weigh 1,900. Brown Swiss are noted for their sturdy ruggedness and long lives, giving milk with high butterfat and protein content

Because of their easygoing dispositions, these cattle are often a favorite for 4-H and youth projects.

## Milking Shorthorns

This breed is red, red and white, white, or roan (a mix of red and white hair). Cows are large, weighing 1,400 to 1,600 pounds; bulls weigh 2,000 pounds or more. They are versatile (first used for both milk and meat) and hardy, noted for long lives and easy calving. Their milk is richer than that of Holsteins, but is not as high in butterfat as that of Jerseys or Guernseys.

## Purebred or Grade?

Do you want a purebred or a grade animal? A registered calf may be of value to you as your herd grows, if you are going to raise a few heifers from her to sell. But don't select a calf just because she's registered. Registration papers will not ensure that your calf will grow up to be the best cow. They won't guarantee high production or good conformation. You are better off buying a good grade heifer than a poor registered one.

Select the type of calf you want in your future herd, making sure she is a good one, whether grade or registered.

You can use a grade heifer as a 4-H project and show her at the fair. But for a dairy show she must be a registered purebred.

## How Do I Choose?

When you have identified your goals, determined which breed, and decided on a purebred or grade calf, you are ready to select your heifer. Use pedigree information and physical appearance to make your choice.

## Pedigree Information

You can use pedigree information to help you choose a good heifer. A pedigree is a record of the calf's ancestry. It gives genetic and performance information about how well the cows that were the calf's ancestors milked. Information about the sire and dam can help predict how the heifer will milk as a cow. If this is the first time you've looked at dairy cow pedigrees and performance information, have someone explain them.

Even though pedigree information is important, common sense and good judgment are just as critical when selecting a heifer. Careful consideration should always be given to the physical appearance of the calf herself.

## Physical Appearance

Look closely at the way a heifer is built and the functional traits that will help determine whether or not she will develop into a good cow. She should have outstanding breed character, which means how closely she resembles her ideal "breed type."

You want an alert heifer with good length of body; a deep, wide rib cage; a long, graceful neck; sharp withers; and a straight back (not humped up or swayed down) with wide, strong loins. Her rump should be level and square, not tipped up at the rear or slanted downward. If her rump is tipped up and her tail head is too high, she will have more trouble calving.

Her hind legs should be straight, not too close together at the hocks or splay footed, and set squarely under her body. Front legs should be straight. She should walk smoothly, without throwing her feet out to the side or swinging them inward. Avoid calves with

## Take Your Time

Don't be in a big hurry to buy a calf. Look at several, so you can make a wise choice. Ask questions. Find out all you can about the calf and her parents.

## Milk Factory

A top-producing dairy cow gives enough milk in one day to supply the average family for a month. The average milk cow will produce 6 gallons a day (96 glasses of milk). A world-record dairy cow can produce 60,000 pounds of milk a year. That's 120,000 glasses of milk!

coarse, flat-topped shoulders or low, saggy backs. She should be well balanced and well proportioned in all of her body parts, not short bodied, shallow bodied, or too short legged.

She should have a good udder. The teats should be well spaced. Future teat size can be determined by the size and shape they have as a baby. You don't want a heifer with really long or fat teats. Udder shape and size when she grows up are difficult to determine; it helps if you can see what her mother's udder looks like, and maybe a photo of her sire's mother.

*Splay footed. Toes pointing out.*

## Age

A newborn calf is cheaper but more risky than an older one. Newborn calves are more likely to get sick unless you buy them from a well-managed dairy and take very good care of them through their first weeks of life. The future conformation of a heifer is more difficult to predict in the newborn calf.

A "started" calf — one that is several months old — is easier to judge, and also less apt to get sick. The easiest calf to take care of is one that is several months old and already weaned, needing no nursing bottles, and past the critical age for scours. But this started calf will also be more expensive.

## Make Sure the Calf Is Healthy

Be sure the calf is healthy before you bring her home. The calf should look bright and perky, lively and energetic, with a glossy hair coat and a sparkle to her eyes. Her bowel movements should be firm but soft, not hard pebbles or excessively runny.

If the calf is dull or slow-moving, has a dull or rough hair coat, foul-smelling manure, or droopy ears, she is sick. Other signs to beware of are a calf that stands with her back humped up, or has a cough or a runny, snotty nose. If you are in doubt as to the health of a calf, have someone else look at her. An experienced person is always a help when you are selecting a calf.

## Registering Your Purebred Calf

If you are buying a purebred dairy heifer or a calf that is already registered, check with your breed association about junior membership. Belonging to a breed association has advantages. The association can provide you with educational material and information and can also help you market your heifer later if you decide to sell her. Register or transfer the registration of your new heifer to your name. Be sure the color markings or the ear tattoo on the registration certificate are correct. For the Ayrshire, Guernsey, and Holstein breeds, you may use photos of both sides of your calf or sketches of the markings. For the solid-color breeds such as Jersey and Brown Swiss, you need an ear tattoo.

# Care of the Young Dairy Calf

**B**efore you bring your new dairy calf home, make sure you have a good place for her. All calves need shelter, but the brand new calf is the most fragile, and needs to be kept warm and dry.

## Fix a Place for Your Calf

One of the ways you can prevent stress and disease is by having a good place for your calf, with adequate shelter that is clean and dry. The calf should have shade in summer to avoid heat stress.

## Shelter

In a mild climate your calf may need only a small three-sided shed, or a protected fence corner with boards or plywood on the sides for a windbreak and a roof over it. Add some clean straw or wood chips for your calf to lie in. But if the weather is cold, you'll need a warm barn stall or even a place in your garage or back porch for awhile, until the baby is several days old and able to live in an outdoor shed.

The calf should always have dry bedding. Moist, dirty bedding contains harmful bacteria and also conducts warmth away from the calf's body. If she has to lie on damp bedding, she may chill. Ammonia gases put off by bedding that is wet from urine and manure

In mild weather, a protected fence corner with clean straw
or wood chips will be adequate protection. But in the colder weather,
your calf will need a warm stall or shed.

can irritate and weaken the calf's lungs and allow
bacteria to become established, leading to pneumonia.

Keep each young calf in a pen by herself, to prevent
spread of disease. Individual calf hutches work nicely.
The hutch is actually a small shed with an attached
outside pen. The calf has her own little barn and a yard
next to it for exercise and sunshine.

A calf hutch is a 4-by-6- or 4-by-8-foot board pen
with a roof. You can hang a water bucket and feed tub
from the wall, about 20 inches off the floor or ground
to keep the calf from stepping in them or getting
manure in them.

## Feeding the Newborn Calf

Choosing a newborn calf from a local dairy gives you the
advantage of being able to take care of her from the start.

### Colostrum

Your calf should have a big drink of colostrum. Some
dairymen let the calf stay with her mother until she
has nursed once or twice. Others prefer to take the calf

away before she has nursed, put her in a clean pen, and feed her with a bottle. The colostrum from the cow is milked out and saved for feeding calves.

When bringing the newborn calf home from a dairy, ask if you can buy a gallon or two of fresh colostrum to take home with you. The extra colostrum can be kept in your refrigerator and fed as long as it lasts. Use very clean containers for storing it.

## Keep Feeding Equipment Clean

Always carefully wash your bottle or nipple bucket after each feeding, or bacteria will grow on it and could make the calf sick. Nipple buckets must be taken apart and cleaned. Use a bottle brush to thoroughly clean a bottle.

## Teaching the Calf to Nurse a Bottle

If the calf was with her mother awhile before you got her, she knows how to nurse from a cow but not a bottle. Your first task is to teach her to nurse a bottle, and quickly. You don't want her to go hungry very long. A young calf should be fed several times a day.

If a very young calf's first few feedings with a bottle are colostrum instead of milk replacer, she'll be more willing to suck the bottle. That's because she likes the taste more. It's also the best food for her at this time.

The keys to getting a stubborn calf to learn how to suck a bottle are persistence, using real milk (preferably colostrum), and having it warm. Young calves hate cold milk. Warm the milk enough so that it feels pleasantly warm on your skin but not hot. If it is too hot, it will burn the calf's tender mouth and she won't suck.

## When to Feed

For the first few days it's best to split the daily feeding into three portions and feed every eight hours. You can feed the calf early in the morning when you first get up, again in the middle of the day, and then the last

## Spoiled by Mama

It's easier to teach a calf to drink from a bottle if she has never nursed from her mother. A hungry newborn calf will eagerly suck a bottle for her first meal. But the calf that has already nursed from her mother is "spoiled," preferring the taste and feel of her udder. She wants Mama! These calves can be stubborn, and it takes patience to get them nursing a bottle.

feeding at night just before you go to bed. Once the calf is a week old you can go to twice a day (every 12 hours, morning and evening), which will be easier.

## How to Bottle Feed

To feed the calf, back her into a corner so she can't get away from you or wiggle around too much. Straddle her neck with your legs if you are tall enough, and you can hold her still with your legs, leaving both hands free to handle her head and the bottle.

Use a nipple that flows freely when the calf is sucking, so she won't have to work too hard at it and get discouraged. But it shouldn't flow so fast that it chokes her. Hold the bottle so the milk will flow to the nipple. The calf shouldn't be sucking air. Don't let her pull the nipple off the bottle.

### Caution

Never overheat milk or milk replacer. Overheating damages the proteins.

A newborn dairy calf is more easily fed from a bottle if she never nursed from her mother.

## Colostrum, Whole Milk, Milk Replacer

If you are able to get some colostrum for your very young calf, divide it into several feedings to get through the first day or two while you are teaching the calf to nurse a bottle. If there is no way to obtain colostrum, use whole milk, preferably raw milk from a dairy, not pasteurized milk from a store. The calf will like the taste of whole milk better than milk replacer.

Before you run out of colostrum or milk, start mixing it with milk replacer if that's what you'll be using, to gradually adjust the calf to the taste of what she'll be drinking from now on. If you make the mistake of switching suddenly to milk replacer, she may hate the taste of the new stuff and be stubborn about accepting it.

## How Much Should I Feed?

It's just as bad to overfeed a calf as to underfeed her. Too much milk can upset her digestion and give her diarrhea. Feed your calf according to her size. A big calf needs more than a little one. Weigh or measure the milk to make sure you are not overfeeding the calf.

Feed 1 pound (about a pint) of milk daily for each 10 pounds of body weight. A calf that weighs 90 pounds should get a total of 9 pints daily — which would be 4½ pints (just over 2 quarts) in the morning and again in the evening. The 90-pound calf would be getting about a gallon a day.

Feed at the same time each day on a regular schedule. This way you won't upset the calf's digestive system. And if the calf starts to get diarrhea from being overfed, immediately cut the amount of milk in half for the next feeding. Then gradually increase it to the recommended amount for her size. As she grows, you can increase the amount of milk, but don't feed more than 12 pounds (1½ gallons) of milk daily.

### Don't Overfeed

Don't feed greedy calves as much as they want; stick to recommended amounts. Make changes gradually. A sudden switch in the quantity or quality of milk can cause digestive problems.

## Calf Feeding Program

| Age | Ration |
| --- | --- |
| Birth to 3 days | colostrum. |
| 4 days to 3 weeks | whole milk or replacer; grain mix or starter. |
| 3 to 8 weeks | whole milk or replacer; grain mix or starter, with access to good roughage. |
| 8 weeks to 4 months | 2 to 5 pounds of calf ration (grain mix) with access to good roughage; calves can be weaned as early as 8 weeks but do better if weaned a little later. |
| 4 to 12 months | 3 to 5 pounds of calf ration with access to good roughage. |

## Feeding from a Nipple Bucket

Once your calf learns how to suck a bottle and knows about nipples, you can feed your calf on a nipple bucket. A nipple bucket can be hung from her fence or stall wall. The advantage of a nipple bucket is that you don't have to hang onto it while she nurses. If you are feeding a group of calves, this can be timesaving. If you use a nipple bucket, hang it a little higher than her head, so she can reach easily.

A nipple bucket saves you time because you don't have to hold it while the calf drinks.

If you use a nipple bucket, don't enlarge the nipple hole. Some people widen it so the milk flows faster, to decrease the time it takes the calf to drink the milk. But this is not a good idea. If the milk runs too fast, the calf may breathe some of it "down the wrong pipe" because she can't swallow it fast enough. This could lead to aspiration pneumonia, which is caused by milk getting into the lungs. The irritation and infection of aspiration pneumonia can't be cured with antibiotics and will kill the calf.

## Feeding from a Pail

You can teach your calf to drink from a pail instead of a nipple bucket. To teach her to drink this way, put fresh warm milk in a clean pail and back the calf into a corner. Straddle her neck and put two fingers into her mouth. While she is sucking your fingers, gently push the calf's head down so her mouth goes into the milk. Spread your fingers so that milk will go into the calf's mouth as she sucks. After several swallows, remove your fingers. Repeat this as often as necessary until she figures out she can suck up the milk. A pail is easier to wash than a nipple bucket or a bottle.

A calf needs your help to drink from a pail.

## Using Milk Replacer

You can buy milk replacers at feed stores. There are many different kinds and brands, and some are better than others. Read the label on the bag to find out what a certain milk replacer contains. If it's confusing, ask a dairyman or your county agent for advice on recommending a good brand.

*Protein and fat content.* The National Research Council (NRC) recommends using one with a minimum of 22 percent protein and 10 percent fat. But calves will do better if the milk replacer contains more fat than that. Milk replacers with 15 to 20 percent fat are better; the calf will grow faster and be less apt to get scours from inadequate nutrition.

*Fiber content.* Check the fiber level in your milk replacer. Low fiber (0.5 percent or less) means it has more high-quality milk products, and not so much "filler."

*Protein sources.* Check the protein sources in a milk replacer. Are they milk-based or vegetable proteins? Milk protein is the highest quality and best for the calf. This is because the newborn calf has a simple stomach. Her rumen is not working yet, for digesting roughages and fiber. She can digest and use protein from milk or milk by-products more easily and efficiently than protein from plants.

*Mixing milk replacer.* Follow the directions on the bag when feeding milk replacers. The powder is mixed with warm water and fed like milk. The recommended amount will vary with different brands.

The powder mixes better if you put the warm water into your container first, and then add the milk replacer to the water and stir well, until it is all dissolved. It won't mix very quickly if the water is cool or lukewarm. Start with water that is a little hotter than you want it to be when you feed the calf, so that it will be just the right temperature by the time you mix in the powder and take it out to feed her.

## Getting Your Calf Started on Solid Feed

A growing calf needs concentrates and roughages. Concentrates are feeds that are low in fiber and high in energy, such as grain. A roughage is a feed that is high in fiber (bulkiness) and low in energy. Hay, grass, corn silage, straw, cornstalks, etc. are roughages. A calf needs roughage to help develop her digestive system so her rumen can begin to function properly.

## Grain

Teach your calf to eat grain or calf starter pellets as soon as possible. Put some in her mouth after each feeding of milk until she learns to like it. Then you can

## Storing Milk Replacer

Keep milk replacers dry and clean. The powder will spoil if it gets damp. Close the bag each time after measuring out the correct amount. Keep it in a container with a tight cover. The quality may be reduced and it can become contaminated with germs if the bag is left open and exposed to light, moisture, flies, and mice.

### Avoid Scours

Scours (diarrhea) in baby calves is not always due to infections. Diarrhea can also be caused by nutritional problems such as overfeeding or use of poor-quality milk replacers.

feed it in a tub or feed box. About 1 cup of grain (¼ pound) is all that a young calf can eat each day at first. Increase it gradually until she is eating about 2 pounds of grain by the time she is three months old.

## Hay

Your calf should have hay as soon as she will start to nibble on it. Calves have small mouths and cannot handle coarse hay, but they will nibble on tender leafy hay. Fine alfalfa, clover, or grass hays — or a mix of these — are very nutritious. Don't give your calf much hay at one time or she will just waste it; baby calves won't eat hay that has been tromped on or laid upon. Give her just a little bit of fresh hay once or twice a day.

## Pasture

Good green pasture is excellent feed for a calf, as long as she is getting some milk (or milk substitute) and grain. She may also need a little alfalfa hay. Make sure she has fresh clean water every day, and access to trace mineral salt. Calves need water, even though they are getting fluid with their milk replacer. Water is especially important in hot weather.

## Flanking a Calf

Dehorning, castrating a bull, and removing extra teats on a heifer should be done while the calf is very young. For these procedures, you may want your calf restrained and lying on the ground. To put your calf down on the ground easily and gently, without having a big struggle, you can flank the calf.

Stand close to the calf, then reach over its back and grab hold of the flank skin with one hand and the front leg (at its knee) with your other hand. Lift the calf off its feet, gently lowering its body to the ground. If you are not tall enough or strong enough to do this easily, have a larger person help you.

To hold the calf still while it's lying on its side being castrated or having its extra teats removed, you can

Flanking a calf is an easy way to lower it to the ground. Lean over its back and take hold of the front leg and flank skin to pick it off its feet, and then set it down gently on the ground on its side.

kneel down and hold the calf's front leg (folded at the knee), putting gentle pressure on its neck with your knee, so the calf cannot rise.

## Removing Extra Teats

Most heifers have just four teats. But some are born with an extra one or two. An extra teat is of no use to a cow and may cause problems when she is being milked. If your calf has an extra teat, it should be removed.

As soon as the heifer is big enough to tell which teat is the extra one, it can be removed, generally when she is one to three weeks old. If you are not sure which one(s) should be taken off, wait until she is older and the teat(s) are more developed, then have your veterinarian do it. Removal of an extra teat in an older heifer should always be done by a vet, for there will be more bleeding, and the wound may need stitches.

The small extra teat on a baby heifer is easily removed. Flank the heifer and put her on the ground. The extra teat is easier to locate when she is lying

### Make Sure

Before removing teats, make sure they are truly extra. You don't want to make a mistake and remove the wrong ones.

This shows how to remove an extra teat with the calf standing. Snip off the teat with clean scissors, making the cut lengthwise with the body.

down. Examine her little udder closely. The four regular teats will be arranged symmetrically, with the two rear teats slightly closer together than the two front ones. An extra teat is usually smaller than the others and located close to the main teats.

After determining which one to remove, disinfect the teat and snip it off with a pair of sharp, disinfected scissors. Hold the scissors with the handle directed toward the front of the calf, and the blades pointed toward her tail end. Make the cut lengthwise with the body. Afterward, dab the severed area with iodine.

## Don't Spoil Your Calf

It's all right to make a pet of your calf, as long as you don't spoil her. Don't let her get away with bad actions such as butting or kicking, or dragging you around when you are trying to lead her. She must respect you. She must learn to have good behavior around people. If she is handled in a gentle manner, but with firmness, she will grow up with a good attitude and be nice to work with.

Another method is to tie the heifer and remove the teat while she is standing. It helps if someone holds her back end so she can't move around. Pull the extra teat down and snip it off cleanly where it meets the udder. Swab with disinfectant. In a young heifer there is rarely any bleeding. Use fly repellent if it's summertime.

## Halter Train Your Heifer

Put a halter on your calf at a young age and teach her to lead and tie up. Train her to lead even if you don't plan to show her. A well-trained heifer is easy to work with for the rest of her life. There are helpful hints in chapter 9 for training a calf. Your young heifer will be even easier to train than a weaned calf because she is not wild or scared.

# Weaning Through Yearling Year

# CHAPTER 15

**A** dairy heifer can be weaned from milk as young as eight weeks old. But a young calf weaned too early, before she is eating enough grain and hay, won't do well. It's better for your calf if you keep her on a nursing program longer.

## When to Wean Your Dairy Calf

How young you wean your heifer will depend on several things, including feed sources, health of your heifer, and how long she has been eating enough solid food to provide all her food requirements.

## Feed Is Important

Dairy calves can be weaned young if they have been eating a special high-quality dry starter ration containing milk products. If the calf will eat enough of this starter, she won't need milk, and early weaning can reduce costs. But don't wean her until she is eating about 2 pounds per day of a grain concentrate mix.

Get the calf eating dry feed as soon as possible. The special grain starter containing milk products can be offered as early as the first week of life, and grain

## Don't Wean Too Soon

It doesn't hurt to keep a calf on milk or milk replacer for quite awhile, but it does hurt to wean a calf too soon. Use your best judgment on when you think your calf is ready to be weaned.

## Get the Rumen Going

Eating grain and hay helps develop the rumen. Very young calves don't have a working rumen. Their digestion is like that of a human, with a simple stomach. They can digest milk, but not roughages. They must learn to eat roughages and get "gut bugs" started in the rumen, to break down and digest the roughage.

should be offered by three weeks of age. The calf won't consume much dry feed at first, but should learn how to eat it. Give her all the grain starter she will clean up daily. By the time she is three to four months old this might amount to as much as 4 to 5 pounds of grain daily. She should also be getting some good hay, starting at a very young age.

If a calf is fed a complete starter containing both grain and roughages, she can be fed as much as she wants, with feed available at all times. She won't need hay until she is about three months old. Grain starter or complete starter can be fed to a calf until four months of age; it will help her through the weaning process. Calves should be eating at least 1 pound of starter daily for every 100 pounds of body weight before they are weaned. Use a weight tape to estimate your calf's weight.

Calves being fed a complete starter don't need hay. Before you discontinue the starter, give some hay for at least two or three weeks. Make sure your calf is eating the hay well. Calves being fed a grain starter should be given hay in addition, and should be eating hay well for at least a week before weaning.

If you can get calves to eat hay at an early age, the rumen will start functioning. Baby beef calves begin eating hay or grass at just a few days of age, following their mothers' example. But the dairy calf doesn't have her mother to show her how to eat. You have to be the substitute mother, encouraging her to eat hay and grain. Put a little grain or leafy alfalfa hay into her mouth after every milk feeding, until she learns to like it.

### The Weaning Process

When you wean the calf, it's easier on her if it is a gradual process. Start by decreasing the amount in her twice-daily bottle or bucket feedings. Cut back to about three-quarters of what you've been giving. Do that for a few days, and encourage her to eat more grain, feeding it right after she finishes her bottle or bucket. She'll then be interested in the grain and not so mad at you for shortchanging her on milk. Then go to one feeding of milk a day, giving grain at her other feeding time. Then stop the milk feedings. Give grain at the time of day you used to offer the milk.

It takes awhile for the rumen to enlarge so the calf can eat enough solid food to give her the nutrition she needs. Right after weaning, young calves still don't have much rumen capacity. They may eat just a small amount of hay compared with the amount of grain they can handle. This will change as the rumen develops more.

## Keep Calves Separate Until After Weaning

If you have more than one calf, they can live together after weaning. It is not a good idea to have dairy heifers with other calves before weaning. Calves like to suck on each other after their milk feedings. Even though they are getting enough from your bottle or bucket feeding, they drink it up quickly and want more. Right after you take the bottle away or the bucket is empty, the calves want to keep sucking. So they turn to each other and suck on each other's ears, or even on the other calf's little udder.

There are problems with calves sucking on each other. In cold weather a calf's ears that are wet from being sucked on by a companion may freeze. Then the ends of the ears will fall off. Even if weather is not cold, you don't want calves sucking each other's udders. This can damage the tiny teats, and can introduce bacteria from the calf's mouth into the teat, causing infection that could ruin that quarter of the udder when the heifer grows up and starts to produce milk.

You can put heifers together about two weeks after being weaned, after they no longer want to nurse one another. If a heifer does start sucking other heifers, she should be kept by herself for awhile.

## From Weaning Until Six Months

After weaning, young heifers can be together in small groups, preferably no more than five to a pen, or out in a good pasture with shelter and shade.

## Feeding Hay

Keep some hay in front of your calf all the time in an area she can easily reach. The hay should be fine-stemmed and leafy, with no mold or dust. As she grows, hay can become a larger part of her diet. After she is five or six months old, good pasture can be used.

### Preventing Infection

Keep pens clean and well bedded. It is very important that heifers do not lie on dirty bedding or in mud and manure. Heifer mastitis and "blind quarters" can result from bacteria entering the teat when heifers have to lie in dirty places. A "blind quarter" is a quarter of the udder that does not produce milk because it has been damaged by infection. If you ever have a heifer with an udder infection, talk to your veterinarian. You should never let this type of infection go untreated, or it may ruin your heifer for milking.

### Feeding After Weaning

After your calf is three months old, you can gradually change from feeding starter to a growing ration. A growing ration should contain at least 15 to 18 percent protein. Often the necessary protein can be supplied by good pasture or alfalfa hay and she'll just need some grain to supplement it — about 4 to 5 pounds of grain daily.

### Feeding and Management Until Breeding Time

Dairy heifers are usually bred at 15 months of age. This is the goal to shoot for, but the important thing is to have your heifer well grown and big enough. Skeletal size and weight are more important than age. If a heifer is not quite big enough at 15 months of age, it is better to wait until she is of proper size for breeding.

### Breeding Weight and Size

A guideline for size at breeding is for the heifer to weigh about 60 percent of her desired mature weight by the time she is bred. This means large-breed heifers (such as Holsteins, Shorthorns, or Brown Swiss) should weigh 800 to 875 pounds at breeding time. For the smaller breeds, this weight would of course be smaller.

*Don't overfeed.* In spite of the fact that you want your heifer to grow well, you don't want to overfeed her on grain. It may make her too fat before she gets her full growth. Her actual skeletal growth may be inadequate with that kind of feeding. If you overfeed her on grain, or shortchange her on protein from alfalfa hay or green pasture, she may have adequate weight but not adequate size. In other words, her weight is due to too much fat instead of body growth.

If a large-breed heifer weighs 800 to 850 pounds as early as 11 to 12 months of age, she is too fat. She has reached breeding weight, but not breeding size. She doesn't have enough height and bone growth, especially in the pelvic area, and will have trouble calving if bred at this time. Wait until she is 14 to 15 months old.

*Don't underfeed.* Underfeeding (lack of feed, or poor-quality feeds) is just as bad as overfeeding. An

## Fat Is Not Healthy

Extremely fat heifers may not become pregnant when bred. In addition, milk-producing tissue will be reduced when fat is deposited in the developing udder; the fat takes up space that would otherwise have become milk glands.

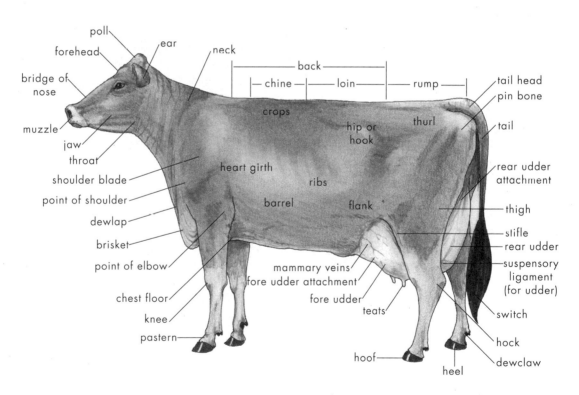

Parts of a dairy cow

## Guideline for Weights and Heights of Dairy Heifers

(height is measured from ground to top of withers)

| Age in Months | Large Breeds (Holstein, etc.) | | Small Breeds (Jersey) | |
|---|---|---|---|---|
| | Weight (pounds) | Height (inches) | Weight (pounds) | Height (inches) |
| 0 | 96 | 29 | 55 | 26 |
| 2 | 170 | 34 | 115 | 30 |
| 4 | 270 | 39 | 195 | 34 |
| 6 | 370 | 44 | 275 | 39 |
| 8 | 500 | 46 | 385 | 41 |
| 10 | 600 | 48 | 460 | 43 |
| 12 | 700 | 50 | 520 | 44 |
| 14 | 800 | 51 | 575 | 45 |
| 16 | 900 | 52 | 650 | 46 |
| 18 | 990 | 53 | 730 | 47 |
| 20 | 1,050 | 54 | 800 | 48 |
| 22 | 1,175 | 55 | 875 | 50 |
| 24 | 1,300 | 56 | 960 | 51 |

undersized, thin heifer may have difficulty calving, produce less milk, and require more feed than a normal heifer during her milking period, because she is trying to catch up to the size she should be.

*Use a weight tape.* You can measure your heifer's height and weight to see how she is growing. You may be able to get a weight tape cheaply or even free at a feed store. To estimate your heifer's weight, put the tape around her body at the girth, directly behind the front legs. Make sure she is standing squarely on her feet. Fit the tape snugly without any slack in it, but not tightly. Then read the measurement that tells her weight.

If you can't find a weight tape, use a flexible cloth tape measure or a string to determine the distance around her girth. If you use a string, figure out the distance by marking the string and measuring it with a yardstick. Then compare this measurement with the growth chart on the next page to see if your heifer is close to what her weight should be for her age.

## Feeding and Management Until Calving

Good pasture is excellent feed for heifers bigger than 500 pounds. Younger heifers may not have the rumen size yet to eat an adequate amount because lush pasture grass is mostly water, and the small heifer cannot eat enough at once. The water takes up too much space. Give the small heifer 2 to 4 pounds of grain daily, and a mineral supplement while she is on pasture. Pasture can supply most of the feed she needs, but be sure it is good pasture. When pasture starts to get dry or short, you can feed alfalfa hay.

If you plan to show your heifer, keep her in a small pen during the summer and work with her daily. That way you can keep her hair looking nice, with regular grooming, and give her special attention and care. And you can keep her on hay and grain, the type of feed that she will have to eat at the show. Switching her suddenly from green grass to hay and grain could cause a problem.

## Preparing a Heifer for Show

If you plan to show your dairy calf, start training her early. Some of the suggestions in chapter 9 may be

Use a weight tape to estimate your heifer's weight.

## An Easy Age

A yearling heifer won't need much pampering. A three-sided shed is usually adequate. She may get a long coat of hair and look rough during winter, but if she is well fed she will stay healthy and keep growing nicely. Keep her shed well bedded and dry.

# Dairy Heifer Growth Chart

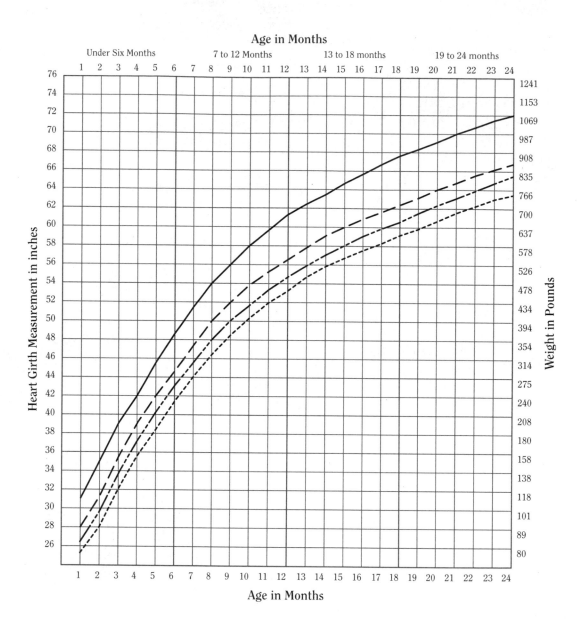

**Age in Months**

Under Six Months | 7 to 12 Months | 13 to 18 months | 19 to 24 months

Heart Girth Measurement in inches

Weight in Pounds

Age in Months

| | Holstein and Brown Swiss |
| --- | --- |
| | Ayrshire |
| | Guerusey |
| | Jersey |

helpful. Use a leather halter during the last part of her training so she can get used to the chain under her jaw. Brush her every day.

## Setting Up

Teach your heifer to stand still and set up. This is a little different for the dairy heifer than the beef animal. The beef calf is always set up "square," with legs positioned perfectly even. But dairy cattle are set up with the hind legs offset; one should be more forward and the other back.

For dairy calves and heifers, the hind leg closest to the judge (or to the camera if you are taking a photo) should be slightly back, and the far leg a little forward. But for cows, the hind leg nearest the judge or camera should be slightly forward. This gives the judge a better view of the back quarters of her udder. For heifers and cows, the front feet should be placed squarely underneath the shoulders.

## Fitting

To keep her looking her best, feed your heifer well and groom her regularly, so her coat has a good shine. When washing her, use lukewarm (not cold) water to soak her thoroughly, then scrub with a good brush. Rinse her with clean lukewarm water to remove all the soap. Don't wash her on a cold day.

Clipping can emphasize her strengths and detract from her faults. You are not trying to fool the judge, but to show your heifer to best advantage. The job you do, whether it is neat or haphazard, shows the judge how you care for your heifer. It reflects the pride you have in her and shows the effort you have made to get her ready for the show. The best way to learn how to clip is to watch an experienced person do it and then practice on your own heifer.

If you are showing in a 4-H class, your club leaders can help you learn how to clip and fit your heifer. They'll give you advice on how to show her,

## Hoof Care

If your calf's feet need to be trimmed, have it done at least a month before the show. Long feet can make a heifer look as though she has weak pasterns or crooked feet because she can't walk or stand squarely.

## Don't Wash Too Often

Washing your calf too often will remove the natural oils from her skin and hair, leaving the hair dull instead of shiny. After you wash her once, she may not need to be washed again except for dirty spots, especially if you keep her in a clean stall with clean bedding. You can also blanket her to help keep her clean.

Brandon Owen grooms his Ayrshire calf as he prepares to take her into the show ring.

and what will be required of you in the show ring. When showing, your main goal is to have your heifer look her best at all times. With patience and practice you can show her in a way that will draw attention to her best points.

## Judging Dairy Cows

If you are going to show your heifer, learn more about judging dairy cattle. You should study the parts of the cow and proper terminology, and the desirable as well as undesirable traits. This can help you in judging, or selecting or culling your own animals. Even though your heifer is not a mature cow yet, you need to know what a good dairy cow should be like. If you belong to a club, take part in the judging contests. They are exciting and fun.

Dairy cows and heifers are all judged the same way, regardless of their breed. They are judged on general

Judge Terry Ziemba looks over the calves shown by Brandon Owen and Sarah Willis.

appearance, dairy character, body capacity, conformation (which includes feet and legs and also the way the body is put together), and mammary system (udder). The judge uses a point system to evaluate the cow or heifer in each category.

## Selling Your Heifer

You may have the opportunity to sell your heifer for a good price. Selling a heifer can be a way to get started in dairying, because the price you get for your heifer will be enough to enable you to buy several more calves.

If you are going to keep her, the next two chapters will tell about breeding, calving, and milking her.

## Get Local Advice

The information in this chapter on feeding your heifer gives you a general idea of how to meet her nutritional needs. For more information on balancing a ration with proper amounts of protein, energy, minerals, etc., and the feeds that might be best for your heifer in the area where you live, talk to your 4-H leader, county Extension agent, or a professional dairyman. The feeds grown in different geographic regions vary greatly; some types of feeds will be more available in your area than others. So get advice from a local professional.

# 16

# Breeding and Calving the Dairy Heifer

**I**f you keep your heifer past her first birthday, you will need to think about getting her bred. A well-grown, healthy dairy heifer can be bred at 14 to 19 months of age. The small breeds, especially Jersey heifers, reach maturity faster than the larger breeds, and can be bred younger and at lighter weights.

## Breeding the Heifer

To freshen at 24 months, your heifer must become pregnant at 15 months. Most heifers will calve about nine months and seven days after breeding.

## Keep a Record of Heat Cycles

When your heifer starts cycling, keep a record of her heat periods, which are usually three weeks apart. That's how you will know when she should be bred. If you know when her last heat period was, you can be watching her closely for the next one, which will come about

## Age and Weight to Breed a Dairy Heifer

| Breed | Age in Months | Approximate Weight |
|---|---|---|
| Jersey | 13 to 17 | 550 |
| Guernsey | 14 to 18 | 650 |
| Ayrshire | 15 to 19 | 700 |
| Brown Swiss | 15 to 19 | 800 |
| Holstein | 15 to 19 | 800 |
| Milking Shorthorn | 15 to 19 | 750 |

17 to 25 days later. If you keep good records, you won't be as likely to miss the period at which you want her bred.

### Selecting a Sire

Breed your dairy heifer to the best bull available. If her calf is a heifer, she will be worth more if sired by a good bull of her breed. If you do not plan to sell the calf as a dairy cow, the sire you choose is not as important. You might decide to breed her to a beef bull that sires small calves. This is one way to make sure she does not have a difficult time with her first calving. A crossbred beef-dairy calf makes a good beef calf if you plan to have him butchered or sell him for beef.

Choose the sire ahead of time and make arrangements with the A.I. technician so you can purchase the semen and have your heifer inseminated at the proper time.

### Make Sure She's Pregnant

Once your heifer is bred, watch her closely for the next few weeks, especially during the time she would have her next heat period. If she does not come into heat at that time, she's probably pregnant. To make sure, have your vet check her for pregnancy two or three months after the breeding.

### Distance Is No Problem

The nice thing about breeding your heifer by A.I. is you can select an outstanding bull from anywhere in the country. You have your pick of the best bulls in the breed. These bulls are kept in central locations called "bull studs." Get a catalog from one of the major bull studs. Your A.I. technician can obtain one for you.

## Preparing for Milking

Get your heifer used to the milking barn or shed at least three or four weeks before she is due to calve. Let her come into the barn and get accustomed to putting her head into the stanchion to eat her grain. If she needs a little supplemental feed, this is a good place to give it to her. She will get used to her surroundings during milking.

## Care of the Bred Heifer

After she is pregnant, keep feeding your heifer for good growth so she will reach her ideal two-year-old weight and size by the time she calves. She will need enough feed for her own growth and the growth of the calf inside her. If she is in good condition at calving time, she will hold up better in milk production. You don't want her thin, or she won't be able to milk as well as she should. And she would also have trouble rebreeding on schedule.

## Prepare a Place for Calving

Prepare a place for her to calve. Make sure it is very clean.

### Pasture

If she is calving in summer and the weather is nice she can calve at pasture, if you check her often. Be sure the pasture is safe and clean, covered with grass, not dirt or mud. There should be shade if the weather is hot, and no gullies or ditches to get stuck in, and strong fences she can't crawl through when she becomes restless during early labor.

### The Barn Stall

If your heifer will have her calf in the barn, you can put lime on the barn floor before covering it with new bedding. The lime not only helps disinfect the floor but also makes a non-slippery base. The floor must have good footing so the calf will be able to stand up, and also so the heifer won't injure herself. She'll be getting up and down during labor, and you don't want her to slip and injure her legs or damage her udder.

The stall should be large and roomy. A stall too small can create problems. If the heifer lies too close to the wall the calf may get jammed into the wall as he emerges.

Make sure the bedding is very clean. Never use wet sawdust, moldy straw, or any damp, moldy, dusty material. Many cases of mastitis are caused by dirty bedding. Wet or dirty bedding containing mold or manure will have germs that can invade the uterus or udder of the calving cow, or infect the calf's navel.

## Signs that Calving Is Near

Start checking your heifer often as she shows signs of being near calving. She will develop a full udder a few days or weeks before calving. Other signs include relaxed muscles around her tail head and vulva, teats filling with milk, or milk leaking from the teats. Every heifer is a little different so just be alert, observant, and ready. If she is with other cattle, put her into a separate pen so they can't bother her when she calves.

## Be There to Help

Many calving problems can be corrected if someone is there to help.

When your heifer goes into labor, check on her frequently to make sure the birth is progressing normally. All too often, assistance is given only after the cow or calf is in critical condition. Be there so you can give help or get help quickly. A normal calf should be born within an hour. Help is most often needed for bull calves, because they are bigger, and for twins. For more details on calving refer back to chapter 11.

## Care of the Newborn Calf

Once the calf is born, the cow should get up soon and start licking him. The licking stimulates his circulation and encourages him to try to get up and nurse. If possible, let her lick him dry. Otherwise, rub him with clean towels if the weather is cold.

## Make Sure the Calf Nurses

Be sure the calf gets a good nursing of colostrum as soon as possible. Wash the cow's udder first so the teats are really clean. Help the calf find the teat, if

### Heat Stress

If the weather is hot when your cow calves, make sure she and the newborn are in the shade. Heat stress can reduce the immunity the calf gets from colostrum, making him more susceptible to diseases.

## Poor Colostrum

Good colostrum is thick and creamy. Thin, watery colostrum is low in antibodies. If a cow does not have good colostrum — if it is bloody or she has mastitis — her newborn calf should be fed colostrum from another cow or from your frozen emergency supply.

necessary. If he doesn't get the job done within an hour after birth, milk out some colostrum and feed it to him.

If the calf was born when you were not there, don't assume that the calf nursed all right just because he is with the cow. It is important you make sure he gets an adequate amount of colostrum.

Without help, most calves will eventually manage to nurse, but some won't, because it is hard for them to get onto teats. This can happen especially if the cow has a low udder or the teats are very full and big. Even if a calf does finally get onto a teat, it may take him too long. His gut lining is starting to thicken after the first couple hours of life, and he cannot absorb enough antibodies.

So be there when your calf is born, and help him nurse immediately, or feed him the colostrum yourself. A 90-pound calf (such as a Holstein) needs about 3 quarts at his first feeding right after birth. A 50-pound calf should have 1½ quarts.

## Care of the Cow After Calving

Offer the cow feed and clean, lukewarm water to drink. She should eat and drink soon after calving. If she won't eat or drink or seems dull, there may be a problem. Consult your vet. Take her temperature to see if she has a fever.

Most cows do fine after calving. Just keep a close watch to make sure all is well. Give her as much good hay as she wants, but only a small amount of grain at first, until all the swelling is out of her udder. If a high-producing cow is fed a lot of grain just before or just after calving, she may have problems.

After calving heifers often have a lot of hard swelling, called "cake," in the udder. It may take several milkings to reduce this swelling, and the udder may be sore until the swelling is gone. Milk her at least twice a day, even if the calf is still with her. He cannot drink all of her milk. This will help reduce the pressure and swelling in her udder, and relieve a quarter the calf may have missed.

It's usually best to leave the calf with his mother for only a few hours. A calf left with the cow more than 12 hours may bunt at her udder when he nurses. This can bruise the mammary tissue, especially if she has a large, full udder or much swelling. Bruising of the mammary tissue could cause mastitis.

## Colostrum and Transitional Milk

The calf can be fed for several days with the colostrum you milk out of the cow. Milk from your heifer cannot be sold, or used by your family for drinking, until it no longer has colostrum in it. But you must milk the cow at least twice a day to get what's left of the colostrum out of her udder and to hasten the production of regular milk. It will take about four to seven days before the milk is ready to be used by people.

True colostrum, the undiluted "first milk," is only obtained from the very first milking or nursing. This is what the calf needs immediately after he is born. This is what you should freeze for later emergencies — the extra first milk that the calf cannot hold. The milkings after that are "transitional milk," a mixture of colostrum and regular milk. It is mostly colostrum at first, becoming more and more diluted by regular milk as time goes by, until there is no more colostrum left in the udder.

The colostrum is thicker and richer than regular milk, and usually quite yellow and sticky. It is waxy when cold. One way to tell if milk still has colostrum is whether or not it goes through a strainer quickly.

# Care of the Dairy Cow

**A**fter your heifer calves, you'll need to milk her. It helps if you know how to milk! Get some practice before she calves.

## Milking a Cow

Milking is easy once you know how. But your arms may get tired the first few times you try. If you've never milked before, visit a friend who has a milk cow, or a 4-H member who has a dairy project.

## Short Fingernails, Clean Hands

When milking a cow, make sure your fingernails aren't too long. If you poke the cow's teat with a sharp fingernail she will not like it! Before starting to milk, you should first wash her udder and your hands, and brush off her udder and flanks if they are dirty. Clean hands and a clean udder will make sure no infection gets into the teats and no dirt into the milk.

## How to Milk

Make yourself comfortable on a stool beside the cow's udder. Hold the clean empty bucket between your legs. Start with front teats or back ones, or the two on the same side if you wish. Hold a teat in each hand and squeeze one at a time, squeezing the milk down and out through the teat opening. Begin the squeeze at the

## Milking Schedule

A cow should be milked twice a day, on a regular schedule (12 hours apart). Not sticking to the schedule can be harmful to the cow.

top of the teat, with your uppermost finger and thumb grip. Finish the squeeze with the lower fingers. By applying pressure with your thumb and index finger, keep the milk from going back up the teat, so when you squeeze with the rest of your fingers, it comes down and out through the hole.

Aim the stream of milk into the bucket or it might squirt off to one side. When you become good at milking, you can direct the stream easily, and can even aim a squirt toward a waiting barn cat. Some cats love to catch a squirt of milk in their mouths.

After each squirt, release your grip. More milk will flow down into the teat. Keep up a nice rhythm by alternating squirts. When those teats are soft and flat, and you can't get any more milk from them, milk out the other two quarters. If one quarter seems to have more milk than another, it means that hand is not yet as strong as your other hand and you didn't get quite as much milk out with each squeeze, taking longer to finish that quarter.

### Sore Muscles

Once you get the proper squeezing motion and rhythm, milking is easy. But it takes muscles in your forearms you may not have used much. They will get tired! You arms may ache if you have to milk very much. But the more often you milk, the stronger they'll get. You'll build up your endurance and the strength of your grip.

### Easy Milkers, Hard Milkers

Some cows are easy to milk. They have teats that are easy to hold onto and they let their milk down freely. The milk almost flows from the udder into your bucket. Other cows are harder to milk. It takes more effort to squeeze out the milk, and longer to milk them. A cow with short teats may be a difficult milker. It is harder to get a good hold. It takes more time to milk if you must squeeze with one or two fingers instead of your whole hand.

To milk, apply pressure with your thumb and index finger while squeezing with your lower fingers.

## Calm and Quiet

When milking, try to keep things quiet and calm so the cow can relax. Feed her some grain to eat while you milk. If she is nervous or upset, she will be tense, and won't let her milk down.

## The Cow's in Charge

Cows can let down or hold up their milk. There is no milk in the teat until the cow lets it down, by relaxing certain muscles that keep the teat canal closed. When a cow wants her calf to nurse, or thinks it is milking time, she lets the milk flow down from the udder into the teat. If a cow doesn't want to let down her milk, you can't get much milk out.

## Prepare Your Heifer for Milking

Always be gentle, quiet, and calm when working around your heifer, so she trusts you and feels secure and at ease. Handle her a lot in the weeks before calving. It will make it easier for both of you when you are training her to be milked.

When you handle her and brush her, touch her udder and teats. Wash them gently a few times. You want her used to being touched, not upset or ticklish. Her udder may be sore when she first calves, because of all the pressure and swelling in it. But if you have handled her udder a lot, she probably won't try to kick you.

### The First Milking

Sometimes a heifer's udder is so sore right after calving that she doesn't want you touching it. She may kick at her calf as you try to help him nurse, or kick at you. If she is just kicking a little because her udder hurts, you can prevent these nuisance kicks by leaning your head into her flank as you milk. Press your head quite firmly against her, if necessary. This keeps her from swinging that leg forward. Your head is pressed into the area right in front of the stifle joint. If you press hard every time you feel her tensing up to kick, she can't kick very well.

If a heifer is nervous the first few times you milk her, talk softly to her or hum a little to reassure her and keep her relaxed.

Sometimes a cow will kick if her teats get sore and cracked, as they may do in cold weather. If they get too cracked, they will bleed. To prevent this, put a little ointment such as Bag Balm on the teats after you finish each milking. This will keep the teats soft and prevent the dryness and cracking that makes them sore.

### Check and Clean the Udder

The cow's udder is a complex structure that needs good care. Before you start milking, check the udder to make sure there are no problems or injuries. Then check for abnormal milk. Before washing the udder, squirt a little milk from each quarter into a small container so you can see the milk easily. If the milk shows any signs of mastitis (lumpy, watery, bloody, or any other abnormality), the cow must be treated with antibiotics.

If the udder and milk are fine, wash the udder with a sanitary solution mixed with warm water. Remove any dirt on the teats so it won't get into the milk. Washing the udder with warm water also stimulates the cow to relax and let down her milk. Use a clean paper towel to wash and dry the udder. Dry the teats before you milk.

## Mastitis

Bacteria sometimes enter the udder through the teat canal. Inside the quarter, they multiply rapidly because of the nice warm environment, and cause an infection. The cow's body sends white blood cells to fight the infection. There is often heat and swelling in that quarter. Infection in the udder is called mastitis. If there's abnormal milk in a quarter, or heat and swelling, it should be treated for mastitis. A cow can look perfectly healthy but still have mastitis.

### Testing for Mastitis

You can test the milk, with a special kit (California Mastitis Testing Kit) that measures the amount of somatic cells in the milk. Somatic cells are white blood cells and mammary cells damaged by the infection. The somatic cell count in the milk will be high as long as there is any infection in that quarter.

Ask your vet or county Extension agent where to get a test kit. Complete instructions come with the kit. It uses a plastic paddle with four shallow cups for samples of milk from each quarter. A special reacting agent is put into each milk sample and stirred with the paddle, creating a change in the milk if there are somatic cells present.

## Treating Mastitis

Your vet has some special antibiotics for mastitis. These are in individual doses. A syringe tube with a long plastic tip is gently inserted into the teat. The medication is squirted up into the quarter. Follow directions on the label for proper use and number of days' treatment. The quarter should be thoroughly milked out before putting in the medication.

## Taking Care of Milk

Take care to keep milk very clean. Use a clean bucket, clean strainer, and clean containers, and wash them all very thoroughly after each use. Rinse all equipment in warm water to remove the milk fat. (Cold water doesn't work very well.) Then scrub everything with hot water and dish-washing soap, using a stiff plastic brush. The brush will clean the equipment much better than a dishcloth. Always use a plastic brush or scouring pad rather than metal. Metal scratches the surfaces and leaves tiny indentations where bacteria can cling. Rinse thoroughly in clean water afterward.

You can pasteurize your milk to make sure it is perfectly safe for drinking. The milk you buy at the grocery store is pasteurized. Pasteurized milk is heated to a specific temperature, kept there for a short time, then quickly cooled.

Home pasteurizing units can be purchased from a dairy supply catalog. These are metal containers with a heating element in the bottom. You fill the container with water, set a gallon of milk in a covered metal pail into the water, then plug in the unit. The water heats to the desired temperature for the proper time, then a buzzer sounds to tell you it is done.

## Feeding the Milking Cow

Proper nutrition is important in determining how much milk your cow will give. She can't produce milk without the building blocks that are provided by good

## Not Safe for Humans

Milk from an infected cow should not be used for humans. Even after the infection is cleared up, the milk can't be used until there is no more antibiotic in it. The label on the medication tells how long this should be. Keep milking the cow regularly in the meantime, to hasten her recovery. In most cases, the milk from a recovering cow can be fed to calves.

nutrition. If you don't feed her enough good feed, she can't produce much milk.

## Hay and Pasture

About two-thirds of the total nutrition of the dairy cow should come from forages — hay or pasture. Make sure the quality is very good. Cows will eat more total volume of hay if you feed it fresh several times a day. The more good forage you can get a milk cow to eat, the more milk she will give.

## Grain

Even though a dairy cow has a large rumen, she cannot eat quite enough forage to supply all her needs. For top production, the dairy cow needs grain. Your 4-H leader or county Extension agent can help you devise a feeding program to create a well-balanced diet for your cow. The amount of grain should be adjusted to fit a cow's needs — more during the peak of her lactation when she is making the most milk, and less toward the end of it.

## Water

Water is an important part of your cow's nutrition, too. She needs a constant supply of clean water. A milking cow needs 3 to 5 gallons of water for every gallon of milk produced. This "water" includes the moisture in her feed. A cow eating hay will need to drink more than a cow on lush green pasture, which contains a lot of moisture.

## Rebreeding Your Cow

Your cow will need to be rebred about three months after she calves. Keep track of her heats when she starts cycling again, so she can be bred at the proper time. It may be several weeks before she starts having heat cycles, but you need to be watching. One clue, if you are milking her, is that she will generally have a temporary drop in milk production the day she is in heat.

After your cow has been bred again, have your vet check her for pregnancy if she does not return to heat. It helps to know if she is pregnant and when she will be due to calve next year, so you can plan the proper time to dry her up before her next calving. If by chance she is not pregnant, you must try to get her rebred.

## Managing the Dry Cow

To be ready for the new calf and new milk production, the cow's body needs a rest from making milk. She should go dry for a couple months before her next calving. Actual length for the dry period can vary depending on the cow's age and condition. She needs at least 45 days' rest to be able to produce a lot of milk during her next lactation, and to make enough antibodies in her colostrum for the next calf. A six- or seven-week dry period is adequate for the average cow. But young cows, such as a first-calf heifer, and high-producing cows generally need eight weeks (56 to 60 days).

When you dry up a cow, do it abruptly. Just stop milking her. It doesn't help to try to ease into it by partial milking. Mother Nature programmed the cow to stop producing milk when her udder is full and tight, which is what happens under natural conditions when a calf is weaned or dies. The pressure in a cow's udder signals her body to stop making milk. She will be uncomfortable for awhile, but after a few days the pressure will ease. Her body gradually resorbs the milk that is left in her udder.

Pasture or hay should provide enough nutrition for her to go through the dry period without becoming fat. She will need grain only if she needs to gain weight.

## Helpful Hints for a Small Dairy

Most of the management concerns of calving time have already been covered. But there are more helpful hints if you have a small dairy.

## Drying Up

The way to help a cow dry up is to reduce her feed, especially grain. Eliminating grain helps her body adjust to not making milk. Check her udder closely while she is drying up. After the last milking, treat each quarter with an antibiotic recommended by your vet to help prevent mastitis. Keep track of her udder as she dries up, to make sure there is no heat or swelling.

## Feeding Colostrum and Waste Milk

If you have several calves, you can save money on milk replacer by feeding the extra colostrum and transitional milk from your cows. Colostrum and transitional milk cannot be sold as milk, but make good food for a calf. Every time a cow calves, her milk for the first five days can be used to feed all your calves, or to mix with milk replacer. The 8 to 25 gallons of colostrum produced by a cow should not be wasted; it will feed calves for quite awhile.

If you feed milk from mastitis cows, feed it only to calves penned individually so they cannot suck on one another. Mastitis germs can be passed from a calf's mouth to another calf's udder if the calves suck on each other.

Many combinations can be used for feeding calves, including mixes of whole milk and milk replacer or sour colostrum and milk replacer. Just be sure to make any changes gradually to avoid digestive upsets. Don't suddenly switch a calf from milk replacer to sour colostrum. Mix it. It's best if you do not change the feed of a young calf during his first two weeks of life. After that the changes won't affect him much, if they are done gradually.

## The Nurse Cow

If you don't need the milk from a cow, she can be a nurse cow. This can increase your income from a dairy cow and eliminate the chore of milking or having to find a way to use the extra milk. With a good nurse cow you can raise four to eight calves each year.

A good crossbred cow (half beef, half dairy) can do well for this, though she may not be able to feed quite as many calves. If the first calf you raise is a crossbred dairy-beef heifer, she could make a very good family milk cow or nurse cow.

A good nurse cow can raise four to eight calves a year.

## Sharing the Milk

You can also share a cow's milk with one or two calves. If you have a dairy cow that gives more milk than your family can use, let her raise her own calf and one extra, while you take part of the milk. You can let the calves nurse one side while you milk the other.

## How Many Calves?

A good mature dairy cow will easily feed four calves at once, but a first-calf heifer may do best after her first calving with just two at a time.

*Raising extra calves on a nurse cow.* Extra calves can be purchased cheaply (see chapter 13) and raised with little effort, since the nurse cow feeds them. She can raise eight big beef steers each year or eight good dairy heifers. A good nurse cow can raise two sets of calves each year, if the calves are fed hay and grain before weaning, so they can be weaned at about four to five months of age. The cow will produce milk for 9 or 10 months before you need to dry her up for her next calf. You can wean the first set of calves and put four more on her. This can be a way to make money for savings or college, if you have room for only one or two cows. Selling 8 to 16 calves a year, raised on one or two cows, is easier than milking them and selling milk, and more profitable.

You need a pen or pens for the calves, next to the milking area. If you are raising dairy heifers, follow the management practices outlined in earlier chapters. You can buy your calves just before the cow is due to freshen, and feed them on bottles until she calves. Or buy them just after she calves. Always keep and freeze a little colostrum if there is extra, to give to any future newborn calves you might buy that have not had adequate colostrum yet.

*Training the nurse calves.* Keep the calves separate from the cow except at nursing time. Get the cow accustomed to eating grain in her stanchion and start the calves on her while she is eating. If she loves her own baby she won't kick it, but she may try to kick off the extra calves. It takes time and patience to convince some cows to accept all those calves nursing at once. Supervise the nursing, and make sure the calves don't bunt her udder too much or get kicked by her.

Let the calves nurse until it is obvious they are done, but don't leave them too long or they will bunt her udder in their greediness for more milk. Bunting calves can bruise the udder. You have to be the referee. If a calf is obnoxious and bunts a lot as he nurses, stand there with a stick and reprimand him every time he bunts, until he learns not to. If she is not hobbled, the cow will kick at him if he bunts, but she may hit the wrong calf.

The hardest job is to get the little buggers back to their pens after they're done nursing. They won't want to leave the cow. You may have to use your stick to encourage them to behave and go back where they belong. You can put halters on the calves before you let them out of their pens. Leading lessons will make them more manageable, as when taking them away from the cow and back to their pen. Double the lead rope over the calf's back while he is nursing the cow, so it is off the ground and won't be stepped on.

## Each Calf Has a Place

Always put the same calf on the same teat. If each calf has his accustomed place at the udder, then it's not such a circus of confusion when you let the calves in with the cow. Be firm and make each calf learn his proper place. If the cow kicks too much, she may discourage a timid calf. You might have to hobble her before you let the calves in, until she learns not to kick.

# Helpful Sources

## State 4-H Programs

Contact your county Extension agent to find the 4-H program located near you.

## Dairy Breed Associations

**American Guernsey Association**
P.O. Box 666
Reynoldsburg, OH 43608-0666
(614-864-2409)

**American Jersey Cattle Club**
6486 E. Main Street
Reynoldsburg, OH 43608-2362
(614-861-3636)

**American Milking Shorthorn Society**
P.O. Box 449
Beloit, WI 53512
(608-365-3332)

**Ayrshire Breeders' Association**
P.O. Box 1608
Brattleboro, VT 05302-1608
(802-254-7460)

**Brown Swiss Cattle Breeders' Association**
Box 1038
Beloit, WI 53511
(608-365-4474)

**Holstein Association of America**
One Holstein Place
Brattleboro, VT 05302-0808
(802-254-4551)

**Red and White Dairy Cattle Association**
Box 63A
Crystal Spring, PA 15536
(814-735-4221)

## Beef Breed Associations

**American Angus Association**
3201 Frederick Blvd.
St Joseph, MO 64506
(816-233-3101)

**Beefmaster Breeders Universal**
6800 Park Ten Blvd.,
 Suite 290 West
San Antonio, TX 78213
(210-732-3132)

**American Brahman Breeders Association**
1313 La Concha Lane
Houston, TX 77054
(713-795-4444)

**International Brangus Breeders Association**
P.O. Box 696020
San Antonio, TX 78269-6020
(512-696-8231)

**American-International Charolais Association**
P.O. Box 20247
Kansas City, MO 64195
(816-464-5977)

**American Chianina Association**
P.O. Box 890
Platte City, MO 64079
(816-431-2808)

**American Dexter Cattle Association**
Rt. 1 Box 378
Concordia, MO 64020-9233
(816-463-7704)

**American Galloway Breeders' Association**
310 W. Spruce
Missoula, MT 59802
(406-728-5719)

**American Gelbvieh Association**
10900 Cover Street
Westminster, CO 80021
(303-465-2333)

**American Hereford Association**
(includes Polled Herefords)
1501 Wyandotte
P.O. Box 014059
Kansas City, MO 64101
(816-842-3757)

**North American Limousin Foundation**
P.O. Box 4467
Englewood, CO 80112
(303-220-1693)

**American Maine Anjou Association**
528 Livestock Exchange
 Building
Kansas City, MO 64102
(816-474-9555)

**American Murray Grey Association**
1912 Clay Street
N. Kansas City, MO 64116
(816-421-1994)

**American Pinzgauer Association**
21555 State Rt. 698
Jenera, OH 45841
(419-326-8711)

**Red Angus Association of America**
4201 1-35 North
Denton, TX 76201
(817-387-3502)

**American Red Poll Association**
P.O. Box 35519
Louisville, KY 40232
(502-635-6540)

**American Romagnola Association**
P.O. Box 450
Navasota, TX 77868-0450
(409-825-8082)

**American Salers Association**
5600 S. Quebec, Suite 220A
Englewood, CO 80111-2208
(303-770-9292)

**Santa Gertrudis Breeders International**
P.O. Box 1257
Kingsville, TX 78363
(512-592-8572)

**American Shorthorn Association**
8288 Hascall Street
Omaha, NE 68124
(402-393-7200)

**American Simmental Association**
One Simmental Way
Bozeman, MT 59715
(406-587-4531)

**American Tarentaise Association**
P.O. Box 34705
N. Kansas City, MO 64116
(816-421-1993)

**Texas Longhorn Breeders Association of America**
2315 N. Main Street, Suite 402
Fort Worth, TX 76106
(817-625-6241)

## Other Organizations

**National 4-H Council**
7100 Connecticut Avenue
Chevy Chase, MD 20815

**National Cattlemen's Association**
P.O. Box 3469
Englewood, CO 80155
(303-694 0305)

**National FFA Organization**
National FFA Center
5632 Mt. Vernon Memorial Hwy.
Alexandria, VA 22309-0160
(703-360-3600)

## Publications

*Working With Dairy Cattle*
(Holstein Foundation, P.O. Box 816, Brattleboro, VT 05302-0816, 800-952-5200)

*Buying, Caring for, Showing Your Angus Heifer*
(American Angus Association, St. Joseph, MO 64506)

*Skills for Life Beef Series*
(Minnesota Extension Service, University of Minnesota, St. Paul, MN 55108-6069)

*Skills for Life Dairy Series*
(Minnesota Extension Service, University of Minnesota)

*Wyoming 4-H Beef Manual*
(College of Agriculture, University of Wyoming, P.O. Box 3354, University Station, Laramie, WY 82071-3354)

*4-H Beef Project*
(PNW 448 — a Pacific Northwest Extension Publication, Oregon State University)

*Raising Dairy Replacements*
(North Central Regional Extension Publication #205, University of Wisconsin, Madison)

*The Calf and Yearling in 4-H Dairying*
(PNW 78 — a Pacific Northwest Extension Publication)

*Feeding and Management of the Dairy Calf: Birth to 6 Months*
(circular ANR-609, Alabama Cooperative Extension Service, Auburn, AL 36849-5612)

*Feeding and Management of Dairy Heifers: 6 Months to Calving*
(circular ANR-632, Alabama Cooperative Extension Service)

*4-H Dairy Cows and Management*
(North Central Regional Extension Publication)

*Kansas Beef Leader Curriculum Notebook*
(State 4-H Office, Kansas State University, Manhattan, KS 66506-3404)

# Glossary

**abomasum (n.)** A compartment of the stomach.

**abortion (n.)** Loss of pregnancy; dead fetus expelled early.

**abscess (n.)** Pus-filled swelling.

**acidosis (n.)** Severe digestive upset caused by too much grain.

**afterbirth (n.)** (Placenta) tissue encasing the calf, attached to the uterus; it comes out after the calf.

**amnion (n.)** Membrane enclosing the calf when he is born.

**antibiotic (n.)** Drug used to combat bacterial infections.

**antibodies (n.)** Protein molecules in the bloodstream that fight a specific disease.

**artificial insemination (A.I.) (n.)** Process in which a technician puts semen from a bull into the cow's uterus, to create pregnancy.

**bacteria (n.)** Tiny one-celled organisms; some cause disease.

**balance (n.)** Harmonious relationship of all body parts.

**balanced ration (n.)** Daily food in the right mixtures and amounts to include all required nutrients.

**Bang's disease (n.)** Brucellosis; causes abortion in cows, and undulant fever in humans.

**birth canal (n.)** Vagina; where the calf comes out from the uterus.

**birth weight (n.)** What the calf weighs when he is born.

**blackleg (n.)** Serious disease of cattle caused by soil bacteria.

**bloat (n.)** Tight, swollen rumen (left side when viewed from behind) caused by accumulation of gas.

**Bos indicus (n.)** Species of humped cattle common to the tropics.

**Bos taurus (n.)** Species of cattle originating in cooler regions.

**bovine (n.)** Term refering to cattle.

**bred (adj.)** Mated, pregnant.

**breed (n.)** Group of animals with the same ancestry and characteristics.

**breeding (n.)** Mating a heifer or cow with a bull. Family history.

**brucellosis (n.)** Bacterial disease (Bang's) that causes abortion.

**bucket calf (n.)** Calf fed milk from a bucket.

**bull (n.)** Uncastrated male of any age.

**bunt (v.)** Hitting with the head (as a calf bunting the cow's udder).

**by-product (n.)** Something made from leftover parts; leather is a by-product of cattle that are butchered for meat.

**calf (n.)** Young animal, either sex, less than a year old.

**calving (v.)** Giving birth to a calf.

**castrate (v.)** To remove the testicles of male cattle.

**cervix (n.)** Opening (or seal) between uterus and vagina.

**cesarean section (n.)** Delivery of a calf through surgery.

**coccidiosis (n.)** Intestinal disease and diarrhea caused by protozoans.

**colostrum (n.)** First milk after a cow calves; contains antibodies that give the calf temporary protection against certain diseases.

**commercial cattle (n.)** Unregistered, not purebred, raised for beef.

**composite (n.)** Uniform group of cattle created by selective crosses.

**conceive (v.)** Become pregnant.

**concentrates (n.)** Feeds low in fiber and high in food value; grains.

**conformation (n.)** General structure and shape of an animal.

**contagious (adj.)** Readily transmitted from one animal to another.

**continental breed (n.)** Originating in Europe rather than the British Isles.

**cow (n.)** Bovine female that has had one or more calves.

**cow hocked (adj.)** Hind legs too close together at the hocks.

**crossbred (n.)** Animal resulting from crossing two or more breeds.

**cud (n.)** Wad of food burped up from the rumen to be rechewed.

**cull (v.)** Eliminate (sell) an animal of low quality from the herd.

**cycling (v.)** Having heat cycles.

**dam (n.)** Mother of the calf.

**dehorn (v.)** Remove horns from an animal.

**dewclaw (n.)** Horny structure on the lower leg, above the hoof.

**dewlap (n.)** Loose skin under the neck.

**digestion (n.)** Process of breaking down feeds into nutrients.

**diphtheria (n.)** Bacterial disease of calves in the mouth and throat.

**disposition (n.)** Temperament and attitude.

**dry period (n.)** The time a cow is not producing milk.

**dual purpose (adj.)** Usable as a breed for both milk and meat.

**electrolytes (n.)** Important body salts.

**EPD (Expected Progeny Difference) (n.)** Estimate of how much better or poorer an animal's offspring will perform compared to the average of all the individuals in the herd or breed.

**estrus (n.)** Heat period; when the cow will accept the bull.

**exotic (n.)** Recently imported breed (European breed).

**feed conversion (n.)** Ratio of pounds of feed eaten to pound of gain.

**feedlot (n.)** Pen where cattle are fattened.

**femininity (n.)** Female characteristics such as udder development, refinement of body, head, neck, etc.

**fertility (n.)** Ability to reproduce.

**fetus (n.)** Developing calf (in the uterus), after 45 days of pregnancy.

**fever (n.)** Body temperature above normal.

**fiber (n.)** Coarse part of the feed, not easily digested.

**finish (v.)** Become fat enough to be butchered.

**fitting (v.)** Clipping, washing, and brushing an animal for show.

**flight zone (n.)** Distance you can get to an animal before it flees.

**foot rot (n.)** Infection (from soil bacteria) causing severe lameness.

**forage (n.)** Pasture and hay; roughages.

**founder (n.)** Inflammation of the hoofs caused by overfeeding grain.

**frame score (n.)** Measure of hip or shoulder height to determine skeletal size.

**freshen (v.)** Give birth to a calf and begin producing milk.

**fungi (n.)** Primitive parastic plants that reproduce by spores.

**genetics (n.)** Qualities and physical characteristics that are inherited.

**gestation (n.)** Length of pregnancy (about 285 days for cattle).

**grade (n.)** Unregistered, not purebred.

**gut (n.)** Digestive tract.

**halter (n.)** Rope or strap looped behind the ears and around the nose to control or lead the animal.

**heat (n.)** (See estrus.)

**heifer (n.)** Young female, before she has a calf.

**heredity (n.)** Transmission of characteristics from parents to offspring.

**hobble (v.)** Tie the legs together.

**hybrid vigor (n.)** The degree to which a crossbred offspring outperforms his purebred parents.

**immunity (n.)** Ability to resist a certain disease.

**infection (n.)** Invasion of body tissues by germs or parasites.

**interest (n.)** Fee paid on money borrowed.

**intramuscular (I.M.) (adj.)** Into the muscle (as an injection).

**intravenous (I.V.) (adj.)** Into a vein.

**iodine (n.)** Harsh chemical used for disinfecting.

**labor (n.)** The cow's efforts in pushing the calf out at birth.

**lactation (n.)** Producing milk.

**legume (n.)** Plant belonging to the pea family (alfalfa, clover).

**leptospirosis (n.)** Bacterial disease that can cause abortion.

**lice (n.)** Tiny external parasites on the skin.

**lump jaw (n.)** Abscess caused by infection in the mouth.

**marbling (n.)** Flecks of fat interspersed in muscle (beef).

**market value (n.)** Price received for an animal.

**mastitis (n.)** Infection and inflammation in the udder.

**maternal traits (n.)** Characteristics that make a good cow.

**muzzle (n.)** Nose and mouth.

**navel (n.)** Area where the umbilical cord was attached.

**omasum (n.)** One of the four stomach compartments.

**open (adj.)** Not pregnant.

**ovary (n.)** Female reproductive gland where eggs are formed.

**parasite (n.)** Organism that lives in or on an animal.

**pasteurize (v.)** To heat milk to a certain temperature to kill germs. Named after Louis Pasteur, who discovered germs and how to kill them with heat.

**pedigree (n.)** Chart of the ancestors of an animal.

**penis (n.)** The male organ that passes sperm into the female, and also passes urine.

**pin bones (n.)** Bony part of the pelvis that protrudes on either side of the rectum.

**pinkeye (n.)** Contagious eye infection spread by face flies.

**placenta (n.)** Afterbirth; attached to the uterus during pregnancy.

**pneumonia (n.)** Infection in the lungs.

**polled (adj.)** Born without horns; naturally hornless.

**production records (n.)** Measure of milk produced, or calf weaning weights, etc.

**progeny (n.)** Offspring.

**protein supplement (n.)** A concentrate containing 32 to 44 percent protein.

**puberty (n.)** Age when the animal matures sexually and can reproduce.

**purebred (n.)** Member of a certain breed (such as Hereford, Angus, etc.). Not to be confused with Thoroughbred (a breed of horse).

**raw milk (n.)** Milk straight from the cow, not pasteurized.

**redwater (n.)** Deadly bacterial disease of cattle.

**registered (v.)** Recorded in the herd book of a breed.

**retained placenta (n.)** The cow fails to shed the placenta quickly.

**reticulum (n.)** One of the cow's four stomachs.

**ringworm (n.)** Fungal infection causing scaly patches of skin.

**roughages (n.)** Feeds high in fiber and low in energy (hay, pasture).

**rumen (n.)** Largest stomach compartment, where roughage is digested.

**ruminant (n.)** Animal that chews its cud and has a four-part stomach.

**scours (n.)** Diarrhea; can be caused by infection or improper feed.

**scrotum (n.)** Sac enclosing the testicles of a bull.

**semen (n.)** Fluid put forth by the bull (containing sperm) when breeding a cow.

**settle (v.)** Become pregnant.

**sheath (n.)** Tube-shaped fold of skin into which the penis retracts.

**show stick (n.)** Long lightweight stick used in positioning a calf's feet and giving him signals when showing.

**sire (n.)** Father of a calf.

**somatic cells (n.)** Cells in the milk, indicating udder infection.

**splay footed (adj.)** Feet toe out.

**steer (n.)** Male bovine, after castration.

**stifle (n.)** Large joint, high on the hind leg, by the flank.

**straightbred (n.)** Animal with parents of the same breed, but not necessarily purebred.

**straw (n.)** Stems of plants grown for grain; often used as bedding.

**stress (n.)** Abnormal or adverse conditions that are hard on an animal.

**subcutaneous (S.Q.) (adj.)** Directly under the skin (as an injection).

**switch (n.)** End of the tail, where the hair is longest.

**tattoo (n.)** Permanent mark in the ear.

**teat (n.)** The "nipple" on each quarter of the udder.

**testes, testicles (n.)** Male reproductive glands.

**udder (n.)** Mammary glands and teats.

**umbilical cord (n.)** Attaches the calf to the placenta and uterus.

**uterine contractions (n.)** Waves in the muscle (like a swallowing motion) that push the calf out of the uterus.

**uterus (n.)** Portion of the reproductive tract where the calf develops.

**vaccination (n.)** Administering a vaccine.

**vaccine (n.)** Fluid containing killed or modified germs, put into an animal's body to stimulate production of antibodies and immunity.

**vagina (n.)** Tube into the uterus from the vulva.

**virus (n.)** Tiny particle that invades cells to cause disease.

**vulva (n.)** External opening of the vagina.

**warble (n.)** Larva of the heel fly; it burrows out through the skin on the cow's back.

**warts (n.)** Skin growths caused by a virus.

**water sac (n.)** Fluid-filled membrane that breaks during birth.

**wean (n.)** To separate a calf from his mother or stop feeding him milk.

**weanling (n.)** Recently weaned calf (up to a year of age).

**yearling (n.)** Calf between one and two years of age.

# Index

Page reference numbers in *italics* indicate illustrations; **bold** indicates charts.